ROBOTS

Some other books by Geoff Simons

Robots in Industry, 1980
Computers in Engineering and Manufacture, 1982
Are Computers Alive? Evolution and New Life Forms, 1983
Introducing Artificial Intelligence, 1984
The Biology of Computer Life, 1985
Is Man a Robot?, 1986
Management Guide to Office Automation, 1986
Evolution of the Intelligent Machine, 1988

ROBOTS

The Quest for Living Machines

GEOFF SIMONS

CASSELL

A **CASSELL** BOOK

First published in the UK 1992
by Cassell
Villiers House
41/47 Strand
LONDON
WC2N 5JE

Distributed in the United States
by Sterling Publishing Co., Inc.
387 Park Avenue South, New York, NY 10016–8810

Distributed in Australia
by Capricorn Link (Australia) Pty Ltd
P.O. Box 665, Lane Cove, NSW 2066

British Library Cataloguing in Publication Data
Simons, G. L. (Geoffrey Leslie), *1939-*
Robots
I. Title
629.892

ISBN 0–304–34086–3

Phototypeset by Intype London
Printed and bound in Great Britain by Mackays of Chatham PLC

Contents

Acknowledgements 7
List of Photographs 8
Introduction 11

1 In Mythology 15
Divine Engineers · Legendary Human Engineers · My Fair Robot · The Prometheus Paradigm · Robots and Religion · Modern Mythologies

2 Early Automata 41
Origins · Further Asian Ingenuity · Towards Mechanical Life · Living Dolls · Mechanical Thought · Enter Electronics

3 Modern Robotics 69
What is a Robot Today? · Artificial Anatomies · Give Us a Hand · The Need to Travel · Motive Forces

4 Intelligent Machines 94
Controlling the System · Aspects of Computer Control · The Need to Know – Senses for Robots · Touching Scenes · The Seeing Machine · Artificial Ears · Smell and Taste

5 The Applications Spectrum 128
In Industry · In Food and Agriculture · In Healthcare · In War · In Prisons and for Security Purposes · In Hazardous Environments · In Training and Instruction · In the Home · In Games and Entertainment

6 Surrogate People 166
The Meaning of Crucial Convergence · The AI Background · Artificial Experts · Advisers and Counsellors · Robot Lovers in Fiction and Film · The Possibility of Real Robot Lovers · Medical Applications · Towards a Robot Anatomy

7 The Impact on Our Attitudes 194
Engineered Anatomies · Cybernetic Systems · Robots and Humans: Programmed for Performance? · Free Will and Creativity

8 Futures 204

References and Bibliography 209
Index 222

Acknowledgements

Thanks are due to Denise Priddey of the Museum of Automata, York, England and to John Richie of the Turing Institute, Glasgow, Scotland for providing photographs and for offering encouragement in the preparation of this book.

Thanks are also due to the following companies and organizations for providing photographs: Advanced Robotics Research Ltd, University of Salford, England; LJ Technical Systems Ltd, Norwich, England; Center for Engineering Design, University of Utah, USA; Hasfield Systems Ltd, Gloucester, England; Tomy Ltd, Sutton, England; British Federal Ltd, Dudley, England; ABB Robotics Ltd, Milton Keynes, England; Staubli Unimation Ltd, Telford, England; and the Ford Motor Company Ltd, England.

I am also grateful to my wife Chris for reading the text in its entirety, for making useful suggestions and for helping in many other ways.

List of Photographs

between pages 96 and 97

1 White clown playing mandolin and black minstrel playing banjo, with moving heads, eyelids and mouths (*Museum of Automata, York, England*)

2 Acrobat clown, able to balance on one hand, tip his body to one side and salute the audience (*Museum of Automata, York, England*)

3 Two biped walking machines, performing at the First International Robot Olympics, September 1990 (*Turing Institute and University of Strathclyde, Glasgow, Scotland*)

4 Friendly robot and admirers at the First International Robot Olympics, September 1990 (*Turing Institute and University of Strathclyde, Glasgow, Scotland*)

5 Robug II: wall-climbing robot performing at the First International Robot Olympics, September 1990 (*Turing Institute and University of Strathclyde, Glasgow, Scotland*)

6 Research robot with stereoscopic vision (*Advanced Robotics Research Ltd, University of Salford, England*)

7 ATLAS II robot with supplementary gripper function and on-board computer brain (*LJ Technical Systems Ltd, Norwich, England*)

8 Utah/MIT Dextrous Hand, a four-fingered end effector jointly developed by the Center for Engineering Design at the University of Utah and the Artificial Intelligence Laboratory at the Massachusetts Institute of Technology (*Center for Engineering Design, University of Utah, USA; photograph: Ed Rosenberger*)

9 Utah/MIT Dextrous Hand, in anthropomorphic applications (*Center for Engineering Design, University of Utah, USA; photograph: Ed Rosenberger*)

10 ARMDROID HS1B robot with standard three-finger gripper (*Hasfield Systems Ltd, Gloucester, England*)

11 Sarcos Dextrous Arm, high performance robot manipulator developed by Sarcos Research Corporation to handle human tools and other objects (*Center for Engineering Design, University of Utah, USA; photographs: Ed Rosenberger*)

between pages 160 and 161

12 Omni 2000 robot for home services and entertainment (*Tomy Ltd, Sutton, England*)

13 Dingbot robot, a toy system ('crazy and fun-loving') that acts capriciously (*Tomy Ltd, Sutton, England*)

14 FEDMAN industrial robot, able to perform a wide range of manufacturing applications (*British Federal Ltd, Dudley, England*)

15 ASEA 2000 industrial robot, able to perform in a wide range of manufacturing applications (*ABB Robotics Ltd, Milton Keynes, England*)

16 Sarcos Dextrous Arm used in conjunction with a force reflective master to form the Sarcos Dextrous Teleoperation System (DTS), enabling a human operator to carry out tasks in a hazardous environment using teleoperated control (*Center for Engineering Design, University of Utah, USA; photographs: Ed Rosenberger*)

17 NEATER (Nuclear Engineered Advanced Telerobot) for automation in the nuclear industry (*Staubil Unimation Ltd, Telford, England*)

18 Industrial robot, used for welding and other applications (*British Federal Ltd, Dudley, England*)

19 Industrial robot with camera vision facility (*British Federal Ltd, Dudley, England*)

20 Industrial robots used for welding on Ford Fiesta manufacturing line (*Ford Motor Company Ltd, England*)

Introduction

It has always been fashionable to describe an age by one of its most important features: so, in history, we have had the Age of Innocence, the Age of Exploration and the Age of Imperialism. In modern times we are often told that we live in the Information Age, the Computer Age and the Global Age; and it is easy to see the weight of such sobriquets. Today we all know that information is a vital resource, as important to national prosperity and national security as oil or gold. It directs government policy, underpins industrial development and sustains military strategy in a hostile environment. Information can thus be recognized as the strongest weapon in a competitive world. Individuals and organizations with control over information have power over citizens, the state and world resources. And the control of information is today intimately linked to the development of computers.

It is possible to depict computation as a central feature of civilization: most social advances have been accompanied by nice developments in the framing and application of numbers, essential ingredients in the ascent of *Homo sapiens*. The conquests of Alexander, the design of the temples on the Acropolis, the advances of Babylonian astronomy, the war devices of Archimedes, the Chinese records inked on silk, the medieval commercial tallies, the dramatic technology of the modern age (including the emergence of the intelligent robot) – all have demanded close attention to number, the ubiquitous discipline of precise quantification. And the very ubiquity of computation has signalled the arrival of the first truly Global Age. Today it is easy to discern a global ecosystem, a global marketplace, a global finance, a global factory and, alas, a global military strategy.

These various potent forces – information, computing and globalization – converge to shape the character of human society and the pace of technological development. Here, for our purposes, we can note the beginnings of the Robot Age, the sturdy seeds from which diverse robot cultures will grow in the decades and centuries to come. Robots are already numerous in factories throughout the world and they are being introduced, albeit hesitantly, into many other human environments – such as schools, hospitals, research laboratories and the home. The word 'robot' is often loosely used: it can denote nothing more than a box of electronic tricks able to automate some trivial task, or it may indicate a highly sophisticated humanoid system equipped, perhaps, with dextrous fingers to deal cards or play the harpsichord. We have always been intrigued most by functional artefacts that resemble human beings or other animate species. In history, when inventive minds were constrained by technological limitations, such artefacts were imagined more often than they were built; or they were contrived as simple mechanical systems, with no access to the electronic magic of modern times.

Chapter 1 charts how the robot dream has existed in the human mind from antiquity to the present day. Fantasy precedes coherent theory, which in turn precedes practical realization: there were many ancient dreams about humanoid artefacts before the earliest engineers and craftsmen began to shape their working models. And we can see how the ancient myths encapsulated the early dilemmas of insecure human beings struggling to cope with an incomprehensible universe. How were people to live? Could they steal secrets from the gods? What was the nature of animation and immortality? Could life be created by magic? It would be left to more modern thinkers to supply more convincing answers. Chapter 2 gives a few examples of the historical automata that first gave substance to the age-old dreams. Engineers across the ancient world, funded by emperors and rich merchants, designed and built humanoid artefacts able to simulate intelligent skills and so to tantalize and mystify gullible observers. Priests used early automata to impress the pious, while entrepreneurs were more interested in making money. The idea of the robot was well established in history: the imaginative fabric of a robot culture awaited – in both dream and nightmare – the enabling technologies of the modern age.

Chapters 3 and 4 profile some of the main features of modern robotic theory and practice. Attention is given to what constitutes a robot in today's mix of convergent technologies: robots are acquiring hands, arms, legs, brains, senses and a range of systems for providing autonomous movement ('animation') in their various environments. And the range of functional robot systems expands by the month. Thus at one extreme the Handy 1 robotic arm, designed by a polio sufferer, Mike Topping, at Keele University in England, serves as an eating aid (*Computer Talk*, 8 October 1990); while at the other extreme an enormous (390 ft/26 m) robot arm – the result of collaboration between companies and institutes in Germany, Britain, Spain, France and Denmark – is able, under computer control, to clean the outside of a jumbo jet in less than three hours; twelve people take eight hours to do the same job (*New Scientist*, 24 November 1990). At the same time robot intelligence expands rapidly as new sensor systems are designed and as further advances are made in the design and fabrication of electronic computation.

A growing range of software is being made available to functional robots, so enhancing their capacity to take decisions, to interact with human beings, to recognize items in their world, and to navigate successfully in a cluttered environment; for example, Mark Reichert and Jed Lengyel of Cornell University in New York are currently developing software to enable a robot to extricate itself from dead ends. Today there is much talk about 'autonomous intelligent machines': papers in the technical press often carry such phrases, and numerous conferences and exhibitions are held on the subject every year. Some 'autonomous intelligent machines' are designed to operate on the earth or in an undersea environment, while others are intended for outer space (thus the six-legged Ambler robot, designed at Carnegie Mellon University at Pittsburgh, Pennsylvania for planetary exploration, is provided with autonomous decision-making abilities by means of a computer sited within a cylindrical body). Modern robots are becoming more agile, more mobile and more intelligent: their claim to function as 'surrogate people' becomes more substantial with every technological advance.

Chapter 5 contains a profile of the extent of robot applications in many 'human environments': in all such cases – in a wide range of service industries, manufacturing industries, research activity and the home – intelligent computer-controlled robots are becoming more numerous, ever more ambitious in their capacity to duplicate human activity in one field after another. Chapter 6 suggests, on a more speculative basis, that robots will be increasingly accepted as 'surrogate people' in a wide range of intimate human functions, able to serve as advisers, counsellors, friends and lovers. For such tasks humanoid robots will rely on such design features as warm, soft bodies; on a wide range of intricate sensor systems; and on a human-level intelligence rooted in knowledge-based systems (the 'expert systems' profiled in Chapter 6). There is already a prodigious and growing technology that will provide the basis for robot applications in such areas. Paralleling the dreams of antiquity, a massive body of modern fiction and film explores how intimate robots may be configured in the years to come.

As robots become more competent, the gap between intelligent artefacts and human beings will inevitably diminish: we will begin to see that robots can be depicted and explored in biological terms, and that human beings themselves can be analysed in robotic terms (Chapter 7). The 'crucial convergence' of technologies that will yield the truly intelligent humanoid robot will be accompanied by an analogous convergence in our attitudes to natural and artificial systems. Such trends will inevitably have profound effects on our view of *Homo sapiens* and on what will be technologically possible in the human/machine societies of the future. Chapter 8 looks at some of the technological trends that will shape tomorrow's intelligent artefacts.

This book is a profile of robot facts, imaginings and speculations. Whole libraries exist on every theme mentioned here, and they are expanding all the time. My aim has been to provide a flavour of the multifaceted historical and modern robot culture, a dynamic matrix that will increasingly influence the development of human society in the years to come.

1 In Mythology

The first robot engineers were the gods of antiquity. The methods that they used were mystery and magic, and the robots that they created were men and women, the first human beings. In such mythologies the first members of the species *Homo sapiens* were obviously artificial systems, designed and fabricated by divinities for reasons best known to themselves. Perhaps the gods were bored and so created tiny homunculi, minute replicas of themselves for amusement or for more serious purposes. Sometimes, the tales tell, the first people were moulded out of clay or earth, or wood was used, to be infused later with spirit or soul. Some deities used a gout of blood or a portion of a man's anatomy – as if, like Mary Shelley's Victor Frankenstein, they had robbed the charnel houses – to build fresh human beings. And there was often a convenient division of labour – one divinity would shape and fashion base materials, while another would then inject an ethereal substance, the essence of soul that would grant human beings their animation and precious immortality.

The early myths posed many important questions, though there were few useful answers. Where did the world come from? In particular, what was the genesis of human beings? How could simple substances be constructed to yield purposeful behaviour and other manifestations of intelligence? Were human beings merely physical systems, or did they carry mysterious components that linked them to supernatural worlds? These and similar questions have perplexed philosophers over the centuries. Only in the modern age, however, have they been realistically addressed by engineers in robotics and artificial intelligence. The early deities, fabricated by human beings, were quick to return the compliment: in a similar way modern engineers are keen to construct artificial systems in man's image.

Divine Engineers

The creation myths, common in many cultures, often have an interest in robotics. When this or that god, according to this or that legend, decided to create the animal world, a vast array of artefacts came into being, artificial systems designed to serve a divine plan that priests were eager to expound. The writer J. J. C. Smart (1959) was one of the first to suggest that Adam and Eve were robots made by God and provided with programs in the form of genes. Here there is no suggestion of souls or spirits, no requirement that intelligence be constructed out of other-worldly stuff. Already it could be seen that the biological 'robots' fashioned by the traditional Judaeo-Christian god could serve as models for new families of artefacts in the rapidly emerging robot age. And Smart's prescient comments were soon to be reflected and developed in countless scientific writings in biology and information science. The biologist Richard Dawkins (1976), for example, was even prepared to suggest that the programs were more important than the biological frames that they inhabited. Here human beings are seen as 'gigantic lumbering robots' that serve as the survival machines for genes. There are no gods in this scenario, no divine engineers; but human beings, even if they cannot be regarded as artefacts, are clearly depicted as physical systems able to display intelligence. It is fashionable to declare that the clever artefacts fashioned by the old divinities have really emerged and evolved in naturalistic ways.

In the ancient legends it is commonplace for gods of various types and with various ambitions to take earth, clay, dust, stone, wood and so on and use such substances to construct men and women. Often, as befits early (and later) sexism, the male is fashioned first and then, as a secondary creature, the woman. There is never any suggestion as to how the gods accomplished their creative tricks: it is inevitable that the creation myths should be collections of evocative one-liners, bereft of technology, but happily conveying the prejudices of the day. Thus, for fertility ritual and other practices, early man shaped clay models, imperfect replicas of human beings and other animals. This suggested to ignorant people how the gods may have set about things.

In the Sumerian legend of Enki and Ninmar the gods created deputies (the first industrial robots?) to perform unwelcome tasks. In this tale, dating to about 2500 BC, the goddess Nammu is said to have fashioned clay into the form of a creature that she desired and then, using her special powers, to have infused the creature with life. Similarly in the Babylonian *Epic of Gilgamesh*, written about 2000 BC, figures shaped out of ordinary substances are animated by divine powers (in some versions of the tale blood is used to provide the necessary animation). In Egyptian myths dating to around 1400 BC the god Khnum modelled a man out of clay, after which the figure was brought to life when the goddess Hather touched it with her staff bearing magical symbols. And the Judaeo-Christian tradition is well acquainted with such fanciful imaginings. Thus 'the Lord God formed man of the dust of the ground, and breathed into his nostrils the breath of life; and man became a living soul' (Genesis 2:7); and, according to the Talmud, dust was collected from around the world, fashioned into human shape and then infused with life. Again it was never made plain how the trick was accomplished, but the magical breath of a divinity often played a part.

For many centuries Swahili gods have been adept at speaking secret words to animate clay models: cultish devotees have related how they can detect blood suddenly flowing through the artefacts to make the muscles ripple and the eyelids move. In Maori mythology the breath of a god is sufficient to animate a human figure built out of sand; and in Melanesia it was thought that the first people were made as wooden puppets by the gods and then brought to life by drumbeats. Some Indian tribes in North America believed that the gods breathed life into models made out of earth and water, but in the legends of the Hopi Indians it was the magical saliva of the Spider Woman that successfully animated the first man and woman. The Okanagon tribe believed that all the races of mankind were created out of earth of different colours. And the Apache god Black Hactin made human beings out of water scum and red ochre, opal and abalone, pollens and white clay. The powerful wind sent by the god to animate the human figures left the imprint of whorls on our fingertips.

In the massive Sanskrit poem the *Mahabharata*, an epic of 3 million words that contains the sacred *Bhagavad Gita*, we encounter Visvakarman, one of the many gods of the mechanical arts that can be found in ancient mythologies. This divine carpenter to the Sanskrit deities is asked to create an artificial woman for the amusement of his brother divinities. Visvakarman combs the world for suitable ingredients ('What are little girls made of?'), and in due course contrives a creature of surpassing beauty. This woman, a seductive humanoid, entrances the gods and stimulates rivalries. Tilottama was perhaps the first of many artificial females capable of arousing sexual desire.

The god Hephaestus, the Greek equivalent of Visvakarman, was said to have built golden maidens to serve the Olympian deities. The artefacts were filled with wisdom, and able to speak and walk; Hephaestus also constructed twenty tripods which 'run by themselves to a meeting of the gods and amaze the company by running home again' (*Iliad*, Book 18). Some tales also tell how Hephaestus, a crippled god, built the giant Talus out of brass to guard Crete by crushing enemies to death against his heated body; and how he used clay to fashion Pandora, the first woman on earth. Daedalus, a descendant of Hephaestus, is said to have created a bronze warrior to confront the Argonauts.

In the myths of the Norsemen the gods specifically designed Midgard, or Mana-heim, to be the abode of human beings, after which it was necessary to people the new domain. Three Norse gods – Odin, Vili and Ve (or, in other versions, Odin, Hoenir and Lodur) – were walking along the seashore when they suddenly encountered two trees, the ash and the elm, roughly shaped to resemble what the gods knew would be the human form. Acting with opportunistic resolve, Odin – in one version of the myth – gave the figures souls, Hoenir bestowed motion and senses, and Lodur provided blood and healthy complexions. The newly created man and woman set about ruling Midgard, peopling it with their descendants, and remaining happy to accept the protection of the gods. (The Greeks, accustomed to shaping figures out of clay, assumed that Prometheus would have done the same in creating mankind; the Northern races, used to making wooden statues, assumed that Ask and Embla, the first man and woman, would be constructed out of ash and elm.)

In ancient China the father-god Yu-ti ('The Ancient One of the Jade') created human beings out of clay, which says something about how things were made in China (the imagined heavenly court was similarly modelled on the imperial court at Peking). In one Assyro-Babylonian myth man was shaped by the goddess Mami from clay mixed with the blood of a slain deity, just as – in a rival tale – human beings were built by the god Marduk out of his own blood. In other mythologies the first human beings are said to have been created out of grass, stones or even excrement. Thus in Oceania in the Pacific the Ata of Mindanao believed the first men to have been manufactured out of local grass; the Igorot suggested that two rushes were employed; and the natives of

Borneo proposed that excrement had been used for the construction of human beings. Elsewhere it is suggested that the first people were carved from stones or from the trunk of a tree, and in some legends it is held that the gods experimented with different substances, not always successfully.

It is, of course, one thing to carve a model of a man or woman, but quite another to give it life. Even existing human beings were often adept at making models and statues, so it was clear that the divine contribution had to be something more – so, in many legends, the gods first shape the figure and then provide animation. Southeast Asia, for example, is a fertile source of such legends.

Spells may be used: the Dairi Bartrak of Sumatra, for instance, are happy to believe that gods can utter incantations that can animate lifeless substances. And again much use is made of the divine breath. The god may simply breathe on the shaped figure, or call up a suitable wind for the purpose. Another option is for the god to journey to heaven to collect the appropriate animating fluid. And there are yet further possibilities. A god, as in Minahassa, may blow powdered ginger at a figure to bring it to life, whereas another god, as in Mindanao, may simply spit on it. According to Bilan myths in Oceania it is sufficient for the god to whip the statue – or to laugh uproariously. Tribes in South Australia believe that the gods formed men from excrement, and then tickled them until they showed signs of life. A god may dance or play drums to provide the necessary animating principle. The gods were nothing if not versatile.

In some myths the first men were produced from eggs – laid by birds or by turtles; alternatively they emerged from egg-shaped objects that had no obvious biological connections. The natives of southeast Borneo believed that human beings derived from mysterious egg-shaped entities in the earth or from egg-shaped foam caused by waves breaking against the rocks. Elsewhere it was thought that the first human beings developed from strange ball-like creatures that were induced by the gods to grow arms and legs. In Tasmania the first people were rounded 'inapertwa', lacking fully shaped limbs, eyes, ears and other elements: the gods intervened to cause the rounded shapes to develop as true human beings.

The differences between men and women were sometimes explained in terms of how they were respectively created. Some tales suggested that the first man was produced by one god, the first woman by another; or that men were shaped out of one substance, women out of another. In the Banks Islands of Oceania the first man was fashioned out of clay, the first woman woven in basketwork; similarly, among Queensland tribes it was held that man was shaped out of stone, woman out of wood. Sometimes the origin of women is seen as too mysterious to be addressed directly: a tribe in the Australian state of Victoria, following the common tale, believed that the first men were formed out of clay, but that the first women were discovered at the bottom of a lake – how they arrived there is not clear. In Papua the first man came from the earth, the first woman from a tree. Among the Baining of New Britain it was thought that stones became men, and birds became women. Another tale tells how God created the first two men, one of whom then set about creating the first woman – out of two coconuts.

In all such legends the gods can work miracles, using special powers to achieve their ends. Sometimes the powers attach to favoured human beings, and they are granted a mysterious creative potential. In the main, however, the ability to produce the first men and women has been considered to be a supernatural talent – a capacity possessed by certain classes of supernatural creatures and denied to ordinary mortals. Therefore it is highly significant when the ancient tales begin to relate how human beings, rather than divinities, can be capable of framing humanoid specimens and giving them life. If the gods were the first robot engineers, then human beings were the next. (Here it is very tempting to digress. The early gods shaped substances into men and women, and then animated them. Human beings later created historical automata and the robots of the modern age. And today it is increasingly common to encounter robots producing robots. A hierarchical chronology of creation over the ages – as each new humanoid family emerges it generates new humanoids of unprecedented types.)

Legendary Human Engineers

It was a mighty step forward when human beings, albeit in fiction and fancy, began to assume divine creative powers. For the first time there was a hint that creative accomplishment could be understood in naturalistic terms, that the shaping of humanoids did not demand divine intercession. Such tales, in existence before the birth of Christ, were the first to suggest that robotics could emerge as a naturalistic science. Even so, as with the Jewish shaping of golems (see below) over the centuries, there was often a need for occult powers, but now tending to be under human control rather than relying solely upon divine initiative.

In the celebrated *Epic of Gesar of Ling*, a prodigious Asian poem, a talented human smith is able to manufacture metal humanoids. At court he requests that the king provide him with ample quantities of gold, silver, copper, iron and bronze. After working alone for only a few days in the palace forge – accomplishing, it seems, a remarkable degree of productivity – he presents to the court the impressive results of his labour. From the gold he has fashioned a life-sized lama and a thousand diminutive monks; the lama is able to move and preach, to the rapt attention of the followers. Out of the bronze the smith has forged no fewer than seven hundred officials and courtiers, as well as a king who is able to discourse on the laws of the realm. From the silver he has made one hundred young girls who sing sweetly; while a general and ten thousand soldiers have been built out of the copper.

The *Epic* mixes occult and naturalistic powers. The smith, a familiar enough occupation, manifestly works in a forge – but there is no detailed description of how he accomplished his wondrous creations. The metallic humanoids are variously dubbed 'magic dolls' and 'metal dolls' – manifestly able to behave like human beings, but operating in ways that are inherently mysterious. The smith is a skilled human practitioner, possessing occult powers but making no claim to divinity. In ancient China some men were thought to possess the skill of *khwai shuh*, the ability to bring humanoid statues to life. This again must be judged a magical talent, the assumption of occult powers that cannot be explained in terms of naturalistic engineering. But at the same time the way was clear for the emergence of engineered automata (see Chapter 2).

The account of the origin of mankind given in the Talmud presents the notion of the golem, the shapeless mass of earth or dust that is later endowed with humanoid form and animation by God or by human beings with occult powers. Again it is interesting to reflect that this myth, persistent in Jewish fantasy, straddles the occult and naturalistic realms. There is a hint here that human beings, albeit with magical powers, can use chemistry (or rather alchemistry – alchemy) to animate lifeless substances. Alchemy, like astrology, occupied for centuries an uneasy position between superstition and science, offering practical methods and techniques but in a confused and contradictory fashion. Gradually science prevailed: the golem, animated by occult powers, gave way to the modern robot, controlled and animated by fluids, electricity and microelectronics, and often carrying an on-board computer.

Much in medieval alchemy (from the Arabic *al-khemia*) derives from the *Cabbala*, a Jewish mystical work. A central aim in alchemy – of particular interest for the purposes of this book – was to bring inanimate substances to life, using a mix of magic and scientific method. Thus the sixteenth-century Swiss alchemist Paracelsus, in a statement that prefigured the modern robotics age, declaimed: 'We shall become like gods . . . we will duplicate God's greatest marvel – the creation of the human.' The alchemists set about their tasks with enthusiasm, but no success whatever – though many claims were made. It was thought prudent to use biological substances: just as the gods of Oceania and elsewhere had a penchant for making human beings out of blood or excrement, so the medieval alchemists were inclined to use such familiar substances as bone, blood, flesh, urine and faeces. The brews were concocted according to magical formulas; there was the nice suggestion that all the procedures were rational and well proven. Alas, the results were far from impressive. Even when semen and female fluids were deployed, to the accompaniment of 'proven' incantations, the inanimate shapes steadfastly refused to burst forth with new life. The alchemists of the Middle Ages would have to wait a few centuries before the birth of animate homunculi in the robot age.

One tale suggested that Simon Magus, using occult powers, had created an animate homunculus as early as the third century. Inspired by such legends, and no doubt by such documents as the prestigious *Cabbala*, later alchemists and necromancers tried hard to repeat the trick. Arnold of Villanova, a thirteenth-century occultist, worked on this task but without success. Cornelius Agrippa had similar ambitions three centuries later, and also tried to create living snakes by inserting the hair of a menstruating woman into human faeces. Paracelsus, keen to assume divine competence, was in no doubt that human beings could be created artificially; in particular, they could be generated 'without woman'. He claimed in fact to have produced 'monstrous dwarfs and other wonderful creatures' using a mix of naturalistic and occult methods. One essential ingredient, he declared, was the semen of a hungry pig, and in *De Generatione Rerum* he explained how this should be used to create a living human being:

> If sperm, enclosed in a sealed glass bottle, is buried in putrefaction of horse dung for some forty days and correctly magnetized, it will begin to live and move. At this time it will bear the shape and form of a human being, but will be transparent and lack body. If now this is nurtured with the arcanum of human blood and kept in the heat of the horse manure it will become a true and living child but smaller than a child of a woman. This is the homunculus. It should be taught with care until it attains intelligence.

Such naturalistic methods seemingly rely little on occult powers, though perhaps a dash of magic might have helped. Paracelsus himself declared of his artificial creatures: 'Through art they receive their life; through art they receive flesh, bones and blood; and by art are they born.' Other men, striving to create life out of inanimate substances, relied less on magnetism and glass bottles, more on the power of spells and incantations. The *Cabbala*, sacred in the Jewish tradition though possibly deriving from earlier sources, relies heavily on particular verbal utterances, anagrams and other specific combinations of letters. It was thought that holy letters could be juggled to create a hot line to celestial power ('In the beginning was the Word').

One central aim in cabbalistic mysticism was to find a combination of words that could be uttered, or written in some ritualistic context, to instil life into a clay figure – clay was much favoured by the cabbalists, as by the ancient gods. As always, the task of creating artificial life – again following the myths of antiquity – consisted in first fashioning the humanoid shape and then infusing it with life by means of incantations or magic brews. Eleazar of Worms, a thirteenth-century cabbalist, declared that specific mystical words had to be recited over each of the figure's limbs and organs to 'breathe life' into it. It was said that a sixteenth-century rabbi, Elijah of Chelm, had successfully animated a golem and been killed by the creature. Most famous was the golem created by Judah Low (sometimes spelt Loew), the Chief Rabbi of Prague, in 1580 to protect the Jewish community from pogroms. The rabbi collected clay in the approved manner and shaped the figure of a man, after which suitable incantations were declaimed and the creature opened its eyes. In one account the golem, by now aged thirteen and possessed of superhuman strength, began to run out of control. Fortunately Rabbi Low managed to pluck the magic formula from the golem's brow, so converting the creature back to harmless clay.

Many of the rabbinical attempts to animate golems relied upon the use of the 'most great name', the mystical 'SHEMHAMPH-ORAS'. A parchment carrying the word could be inserted into the mouth of the golem to bring it to life; conversely the golem could be deactivated by removing (unplugging) the parchment. And other words could be used in the same way. 'EMETH' (Hebrew for 'truth') would activate the golem when written on its forehead (delete the 'E' and the golem would die). It seems that there were problems in controlling golems, once activated: this is a common theme in the creation of artificial life.

The golem tales illustrate an important point. The earliest myths assign virtually all creative powers to the divinities: only the gods can create men and women, can imbue inanimate substances with life. The mystical powers reside in an untouchable supernatural realm, far outside the scope and ambitions of mere human beings. But gradually, perhaps aided by a burgeoning but primitive science (and perhaps also by the theological scepticism of priests), human beings began to gain a new confidence. Perhaps, after all, creative powers could be understood by mere mortals and used by them. Perhaps there were ways of tapping into celestial creativity, of uncovering the occult forces that might or might not have a divine source. Perhaps human beings, without divine assistance, could become true magicians, breathing life and soul into base materials. With science in its infancy, the cabbalists and necromancers were forced to rely upon superstitious methods: alchemy prefigured chemistry and physics, various 'vitalisms' prefigured biology. But already the route was clear: human beings were becoming arrogant enough to assume divine powers. They were on the way to creating the animate artificial systems of the modern age.

There are two particular myths that are immensely potent in our approach to robot systems. In the legends of Pygmalion and Prometheus layer upon layer of significance is encountered, as if the ancient mythmakers could discern all the confusions and paradoxes that modern man would experience in his confrontation with modern machines, particularly with those machines that might be deemed to have intelligence and animation. These two myths deserve separate treatment, however brief.

My Fair Robot

On the island of Cyprus lived Pygmalion, a young sculptor devoted to his art. At the same time he was disgusted by the local girls, the Propoetides, who impiously denied the divinity of Aphrodite. The myth suggests that the goddess resolved to punish the girls by inspiring in them such immodesty that they were eager to prostitute themselves to all and sundry. Pygmalion, evidently a very proper young fellow, was horrified at the women's behaviour and determined to shun their society. Instead of consorting with earthly women he fervently venerated Aphrodite and took comfort from his world of statues.

In due course, Pygmalion produced an exquisite ivory statue of a woman whom he called Galatea. Captivated by her beauty, he wished that she were alive. With this in mind he prayed to Aphrodite who, impressed with the young sculptor's piety, decided to grant his wish. One day, when Pygmalion was embracing his statue, he suddenly felt the ivory turn to flesh, saw the eyes begin to open, and found that his fervent kisses were being returned. With the help of a goddess, Pygmalion had created his perfect woman: it is to be hoped that she lived up to his moral expectations.

The myth highlights the perennial human search for the ideal, the restless yearning for a world bereft of all earthly imperfections. It also focuses attention on how human beings might relate to artefacts. Are we to encompass them to the exclusion of normal human relationships? (In the modern world there has been much discussion of how young people can become addicted to computer-based systems to the detriment of normal psychological development.) Is it possible that humanoid robots will come to serve as surrogate people, programmable to meet individual needs? What does this say about the durability of more traditional human relationships? (These matters are discussed further in Chapter 6.) Galatea was the ancient embodiment of the dream that has persisted into modern times. Are we to love images, stereotypes, façades and replicas rather than real people? In the age of the increasingly competent robot, to what extent will artefacts come to supplant, rather than supplement, normal human relationships?

The Pygmalion myth has been reworked many times, as in Giambattista Basile's *Pentamerone* (1636), in which the sexes are reversed. Here, like Pygmalion, the central character, Bertha, is largely unimpressed with the members of the opposite sex. So she mixes almonds, pearls, scented water, sapphires and sugar, following a special recipe, to produce a beautiful young man. Bertha then makes a suitable prayer to the goddess of love, the figure comes to life and a happy consummation ensues. Similarly, in Oscar Wilde's story of Dorian Gray the portrait comes to life; and we are all familiar with the robot lovers, in modern film and fiction, that serve as surrogate partners for men and women.

Medieval theologians worried constantly that demon lovers (incubi and succubi) were infinitely more seductive than human beings, however sinful. The modern analogue, already reflected in fantasy, is that another non-human category – the humanoid robot – may come to expose the shortcomings of all types of human–human intercourse. Pygmalion, we now see, was his own sort of robot engineer: modern robot engineers, less likely to need divine assistance, are much more likely to achieve dramatic and far-reaching results.

The Prometheus Paradigm

In Greek mythology Prometheus was a Titan, one of a race of gods on Mount Olympus. The Titans had suffered a crushing defeat when they rebelled against the other divinities, though Prometheus, with characteristic cunning, had decided upon a prudent neutrality – and so been permitted to stay on Olympus. Nevertheless he continued to harbour a silent grudge against the destroyers of his race, and was apt to favour human beings over the gods whenever there was a conflict between them. Prometheus in fact had created the human race, out of clay, or earth and water, or perhaps out of his own tears (though even Prometheus, a senior deity in the pantheon, had to rely upon Athene to breathe life and soul into the first man).

The real problems began when the human race, faring badly, was in serious danger of dying out, something that Prometheus could not tolerate. His remedy was to steal fire from heaven and to present it to mankind to aid the survival of the race. In one tale Prometheus journeys to the island of Lemnos, where Hephaestus tended his divine forges. Here he stole a brand of the holy fire, enclosed it in a hollow stalk and brought the sacred gift to mankind. In another version Prometheus used the fire of the sun to light his torch. Zeus responded in anger by asking Hephaestus to create Pandora out of clay, whereupon Hermes planted perfidy in her heart. Pandora then travelled to earth with her magic vase ('Pandora's Box') to release terrible afflictions upon the world.

Zeus also commanded that Prometheus be bound with indestructable chains to one of the crests of Mount Caucasus. There the prisoner would be visited by a winged monster (or an eagle or a vulture) that would pluck out his liver and devour it. Some observers have made much of the fact that it is the liver that manufactures substances essential to the efficacy of blood, the very fluid of life. The Greeks were well aware of such biological details, and so the punishment of Prometheus may be interpreted as a reminder that the gift of life can only be properly given by the gods. But despite his torment Prometheus would not yield. He continued in his defiance of Zeus, expressing his hatred in violent outbursts. Perhaps he knew that, having stolen the secret of fire from the celestial realms, he was a threat to Zeus himself.

Eventually, perhaps through common sense, Zeus relented. After thirty years – some versions of the tale suggest thirty thousand – Hercules was sent to slay the winged monster and to break Prometheus's chains. The torment was over, but the myth endured. One sign of its relevance to the modern age is that Mary Shelley gave her seminal *Frankenstein* the subtitle 'The New Prometheus'.

It is now possible to highlight some of the central features of the Promethean myth, key themes that bear on human efforts to build animate robots in the modern world of science and technology. The myth insists that any human being who tries to build life artificially is acting outside the legitimate human province, carelessly straying into the divine orbit. The implication is that the gods will soon be inspired to vengeance. A further theme is that artificial life will be unsuccessful, lacking the divine spark – spirit or soul – that can render it truly animate in a difficult world. If the gods are not involved in the creation of life, then it is inevitable that the newly shaped creatures will be flawed: man is imperfect and so, necessarily, are all his artificial products. But human beings, it is implied, are not responsible for their imperfections: because they were initially created without divine sanction, but only through an independent whim of Prometheus, the blame must attach to their creator. In modern terms, the flaws in animate robots invite criticism of the human engineers, not of the machines themselves. In essence all attempts to usurp the role of the gods – especially in their creative modes – can only lead to destructions and chaos. This central message runs through many versions of the myth: Prometheus, the golem and *Frankenstein* are only the most obvious examples.

In the modern world such myth-messages are likely to fall on deaf ears. Like the medieval Paracelsus the modern scientist and engineer may aspire to divine powers, except that today the divine has been wished away, consigned to the detritus of prescientific times. What is true in the myth is obvious enough: man, and man alone, is responsible for the flaws in his machines. But where does the fault lie in man? In his physical structure? His spiritual awareness? His moral insight? And, if the flaws are discerned and acknowledged, where is the blame properly placed? Prometheus was, after all, one of the gods, albeit with maverick impulses. Is the divine realm not somewhat lacking in having created imperfect human beings?

We can at best take the ancient myths as metaphors for what is happening in the modern technological world. Human beings were once viewed as artificial creations, analogous in important respects to the modern intelligent robot. It is obviously true that we should think carefully about the world we are creating, but we do not need to imagine that Hephaestus or Prometheus have much to teach us, other than about the tale-telling capacities and conceptual limitations of a less sophisticated age.

Robots and Religion

It has already been seen that there are many – some unexpected – connections between the emergence of artificial robots and elements in the religious attitude. This is not altogether surprising: many of today's robots are made in man's image, as are the gods and all the associated supernatural realms. Robots and gods, both sired in the human mind, are inevitable siblings. And we all know that siblings can have different interests, different preoccupations, different orientations.

The Promethean myth suggests that there should be religious hostility to the spread of robots in the modern world, and there is much supporting evidence for such a view. For example, the Christian hostility to science has been demonstrated in countless ways over many centuries: the early opposition to naturalistic Greek thought, the burning of the famous library at Alexandria, the murder by Christians of the mathematician Hypatia, the persecution of scientific heretics. Galileo, old and nearly blind, was dragged before the Inquisition; Darwin and Freud abused and insulted by ignorant clerics; Bertrand Russell proclaimed unfit to teach mathematical philosophy at the College of the City of New York. There were pious proclamations against steam locomotives, anaesthetics and lightning conductors. When earthquakes occurred in 1755, the Rev. Dr Price attributed the devastation in Massachusetts to the 'iron points invented by the sagacious Mr Franklin'. The worthy cleric was quick to point out: 'In Boston are more erected than elsewhere in New England, and Boston seems to be more dreadfully shaken. Oh, there is no getting out of the mighty hand of God!' The Christian anxiety, reiterated over the centuries, about all scientific matters cannot be gainsaid; but another thread, in Christianity and in other religions, can also be discerned.

It has been suggested that Japan's relative success with robots can be attributed, at least in part, to traditional cultural values in which religious attitudes play a significant part. Thus in 1982 Henry Scott-Stokes wrote in the *New York Times Magazine* of Japan's 'love affair with the robot', observing that it 'dates back several decades and, in the view of many authorities, is a unique, intensely personal reaction with roots in Buddhist values'. Similiarly Koichi Kawamura, writing in the *Oriental Economist* in 1983, declared that it was religion that encouraged his countrymen to embrace robots, proclaiming the Japanese 'mentally constituted to accept in place of the sword, for instance, the computer or the robot as one of their gods if they should come to believe that such things would protect them from external enemies or disasters'.

A remarkable feature of Japanese culture is the attitude to animism that has endured into modern times. Animists believe that many natural objects, alive or not, can have their own *kami*, an in-dwelling soul or spirit. Rocks, hills, trees – all can possess an individual *kami* that makes them worthy of worship; and the same principal can even be applied to man-made objects. Samurai swords can have their own souls, as can industrial tools, including that greatest tool of all – the intelligent industrial robot. The robot pioneer Joseph Engelberger, head of Unimation, has graphically described how the Shinto religion has been deployed in Japan to bring industrial robots within traditional culture. At one factory he witnessed a Shinto ritual consecration of two new Unimate robots: 'Two Shinto priests are there, banging their sticks and moaning and groaning and making all kinds of different sounds, blessing the robots and blessing the general manager and blessing me, with garlands of flowers around the robots.' The general manager then invites the assembled people to welcome the robots, 'your new fellow workers', whereupon everyone claps.

The element of animism in traditional Shinto religion helps the Japanese to make friends with artefacts, particularly when machines have anthropomorphic character. (The same attitude is also well known in the West: here there is much evidence of how human beings – adults as well as children – come to regard their computer-based systems as 'persons'.) At the Fanuc plants at Hino, at Nissan's Zama plant, and at other sites of robot applications, Japanese workers have given the industrial robots the names of popular singers, actors and other celebrities. The Tokyo psychologist Seiichiro Akiyama, quoted by Scott-Stokes, has commented of the robots: 'We give them names . . . We want to stroke them. We respond to them not as machines, but as close-to-human beings.' The point should not, however, be overstated: there are now far too many robots in Japan for them all to merit such singular personalized treatment. None the less it is interesting that there are elements in Shinto and in Buddhism, absent in Christianity, that encourage an enthusiastic attitude to robot culture. For example, the Japanese word *dogu* (meaning 'tool') has a spiritual significance for Buddhism and came to have a general relevance, variously denoting an aid to sanctity, religious clothing, samurai weapons and tools used in production. It was believed that, if a worker could recognize and exploit the magical quality in his tools, he would be more productive. Thus peasants would make offerings to their washed and purified tools, and pray for the preservation of their unique powers. In such a cultural context the Japanese enthusiasm for the modern industrial robot is not hard to understand.

Various observers have commented on the hostile attitudes in the West to the emerging robot culture, and sometimes a religious dimension is perceived. For example the author and scientist Isaac Asimov, robophile rather than robophobe, has commented in the introduction to an anthology of robot science fiction that 'the creation of a robot, a pseudo-human being, by a human inventor is . . . perceived as an imitation of the creation of humanity by God. . . . In societies where God is accepted as the SOLE creator, as in the Judaeo-Christian West, any attempt to imitate him cannot help but be considered blasphemous.' Similarly the mathematician and computer pioneer Norbert Wiener, in his nicely titled *God & Golem, Inc*, felt obliged to address himself to the impact of cybernetics on religion. Here he remarks that just as Rabbi Low of

Prague, in supposedly animating the golem, had incurred the wrath of the pious, so the potential of modern cybernetics is sometimes likened to the sin of sorcery. Any medieval robotmaker would have been roundly condemned – and probably burnt – by the Holy Inquisition. And Wiener had also suggested, in contrast to the comments above, that 'Buddhism, with its hope for Nirvana and a release from the external wheel of Circumstance . . . is inexorably opposed to the idea of progress'. But the Japanese philosopher and roboticist Masahiro Mori has declared that 'to learn the Buddhist way is to perceive oneself as a robot': in his book *The Buddha in the Robot* Mori suggests that 'robots have the buddha-nature within them – that is, the potential for attaining buddha-hood'. In the same vein Sueo Matsubara, the acting president of the Mukta Institute which was founded by Mori in 1970, has commented that 'You really can't make a good robot without chanting the scriptures.' Perhaps Western technologists, struggling to compete against the vitality of the Japanese economy, had better take note.

It is clear that there are multifaceted links between religious speculation and robot theory, between ancient and modern attitudes. What we see is a complex weave in which old patterns are never wholly supplanted by new ones, in which new shapes are influenced by the forms that emerged in earlier times. It is easy to detect the resulting tensions in Western thought. Is life inextricably associated with spirit and soul? If robots and other computer-based systems can take decisions, then what of free will? If, like Mori, we are to see human beings as robots, then what becomes of moral responsibility? (Such topics are addressed more fully in Chapter 7.) The emerging families of animate robots will encourage us to speculate on age-old questions, persistent anxieties about the nature of man and the purpose of life. New technologies have always stimulated the human imagination, leading in turn to further technological advance and yet more dreams. But the endless dialectic between thought and technological accomplishment has taken on a new dimension with the birth of the functional robot. This is because of the anthropomorphic character of the new machines: we find that today's robots have arms, hands, legs, sensitive skins, sense organs and brains. With the brain of the modern robot comes an emerging life, a short remove from spirit and soul. What would the medieval theologians make of this? What do modern theologians think of it all?

Ancient myth was indistinguishable from ancient religion: the fanciful tales attached to gods and goddesses, giants and Titans, who expected worship. Many of the supernatural creatures had their earthly favourites, once the human race – a fresh family of artefacts – had been created; and many other terrestrial penchants were imposed, in imagination, on the celestial realm. Thus the gods were lovers, liars, warriors, schemers, travellers, smiths: their activities had importance for earthly affairs, influencing to fundamental effect the course of human history.

Today there is a different picture. In the West at least, myth has become divorced from religion; myth is powerful in fiction and film, but religion is often perceived as a moribund irrelevance to anything important. Consider the robot-linked tales in modern literature and feature films – religious thinkers rarely address the questions that these raise. What is the significance of the newly emerging robot species in our midst? How are we to relate to artificial creatures that will become as intelligent as ourselves? If artificial life develops to be indistinguishable from human life, apart from the matter of origins, what does this imply for our precious status, the uniqueness of *Homo sapiens* that we have always been eager to assume?

There is no doubt that the intriguing new questions will be richly explored via the modern mythologies, via the fiction and films that focus on the robot theme. It is here, rather than in papal encyclical or Sunday sermon, that such topics will be increasingly discussed. And it is useful to remember that the discussion has already begun: people do not always realize that there is a substantial tradition in this area.

Modern Mythologies

The rapid development of industrial technology in the nineteenth century saw a corresponding evolution of fiction dealing with science in an imaginative way. Long before the time of the modern industrial robot tales were written about humanoid creations, robotic inventions that mimicked human behaviour and variously seduced or terrified the human beings who encountered them. The French tale *L'Eve Nouvelle* (1879), later known as *L'Eve Future*, concerns an attractive artificial woman animated by electricity and possessing a soul. The author of the tale, Villiers de l'Isle Adam, is supposed to have proclaimed: 'My master, Edison, will soon teach you that electricity is as powerful as God.' And another female automaton, called Olympia, is encountered in *Der Sand-*

1 White clown playing mandolin and black minstrel playing banjo, with moving heads, eyelids and mouths

2 Acrobat clown, able to balance on one hand, tip his body to one side and salute the audience

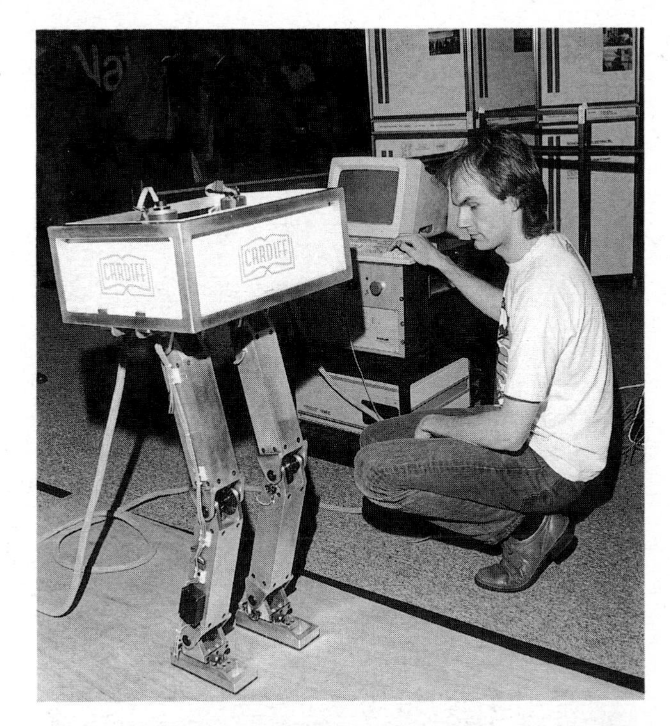

3 Two biped walking machines, performing at the First International Robot Olympics, September 1990

4 Friendly robot and admirers at the First International Robot Olympics, September 1990

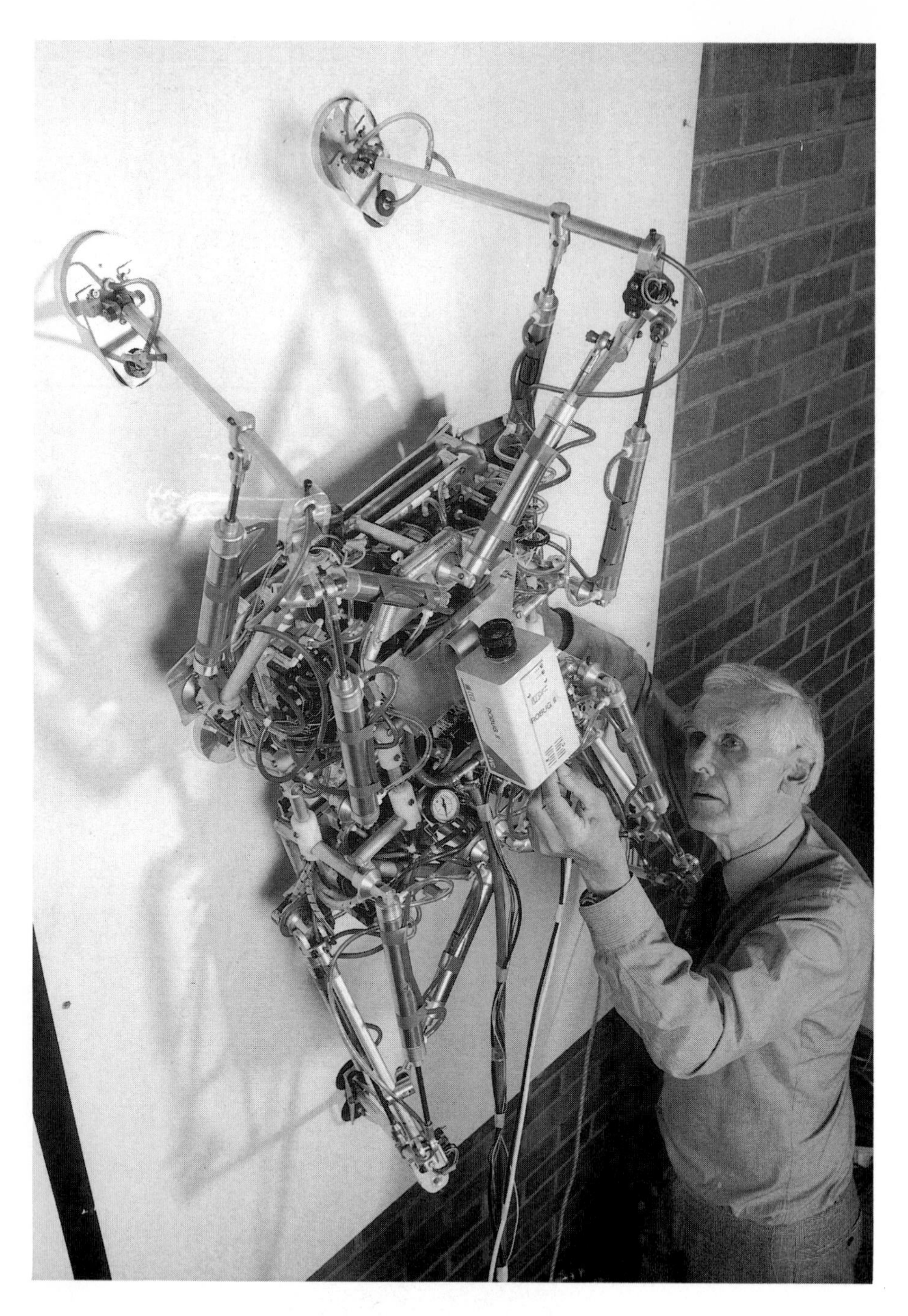

5 Robug II: wall-climbing robot performing at the First International Robot Olympics, September 1990

6 Research robot with stereoscopic vision

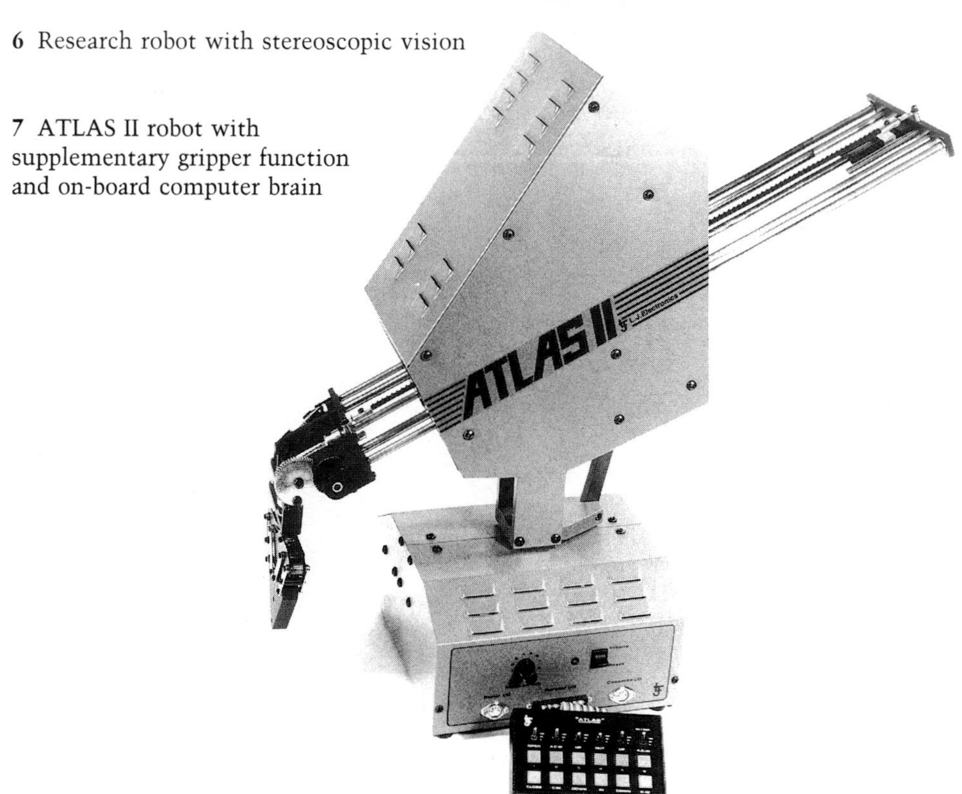

7 ATLAS II robot with supplementary gripper function and on-board computer brain

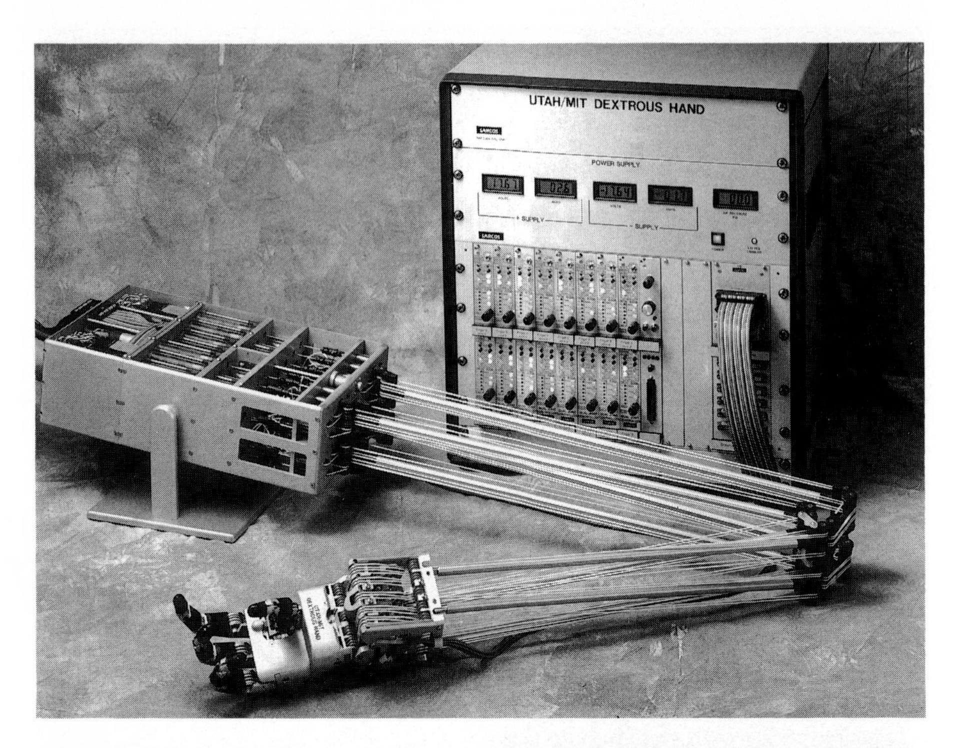

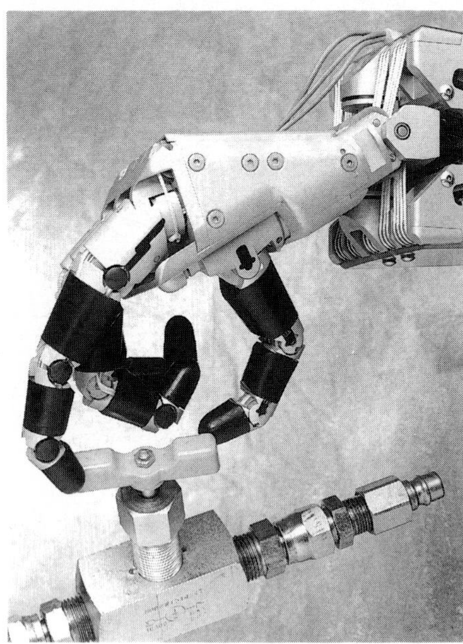

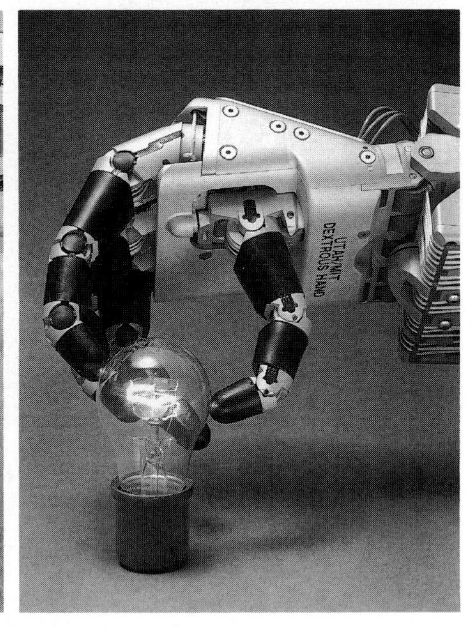

8 Utah/MIT Dextrous Hand, a four-fingered end effector jointly developed by the Center for Engineering Design at the University of Utah and the Artificial Intelligence Laboratory at the Massachusetts Institute of Technology

9 Utah/MIT Dextrous Hand, in anthropomorphic applications

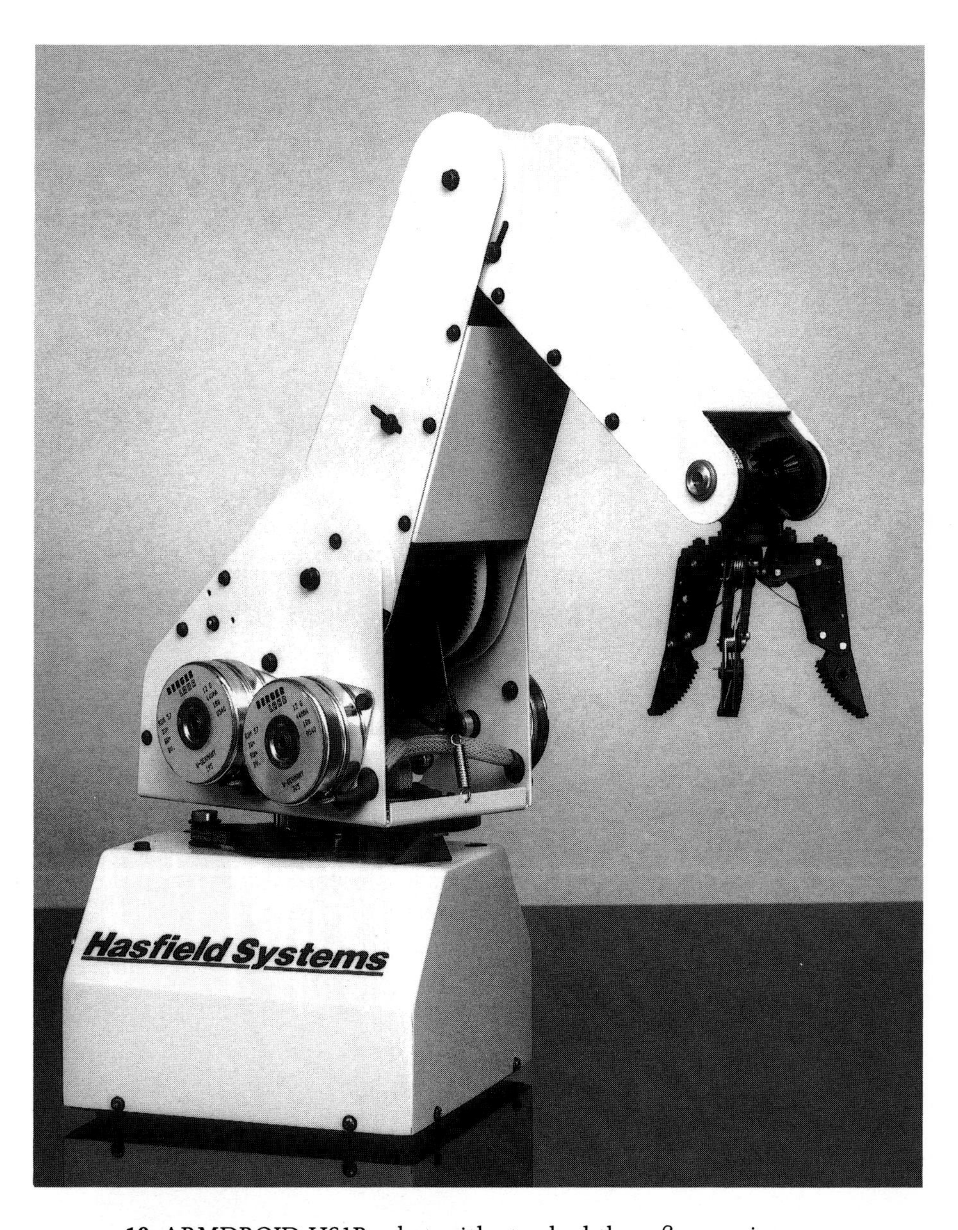

10 ARMDROID HS1B robot with standard three-finger gripper

11 Sarcos Dextrous Arm, high performance robot manipulator developed by Sarcos Research Corporation to handle human tools and other objects

mann by the German writer E. T. A. Hoffmann (1817); she behaves, disturbingly enough, in human fashion, though she can only keep herself wound up by sneezing. Yet another female humanoid is depicted in Lester del Rey's *Helen O'Loy* (1938); she is engaged as a housekeeper to two bachelors and in due course marries one of them.

Perhaps the best-known fictional treatment of the humanoid automaton theme is Karel Capek's *R.U.R. (Rossum's Universal Robots)*, a play first produced at the Prague National Theatre in 1921. (It is customary in modern books to declare that 'robot' derives from the Czech word for 'worker' (*robota*), just as Asimov's Three Laws of Robotics* have been quoted countless times.)

Capek was nothing if not ambitious. He set a vast production plant on a small island, and envisaged the mass production of artificial people. Here he took the obviously biological route: brews are mixed in large vats to form the basis for livers, brains and other robot organs; and prodigious spinning wheels are employed in manufacturing endless miles of veins and nerve fibres. All is well until the robots acquire feelings, at which point they can no longer tolerate their existence as slaves. They rebel and threaten the existence of the human race, in due course promising that the world will be inhabited by artificial beings.

Capek is clearly aware of many of the questions raised by the old mythmakers. What happens when machines become conscious? Is it possible to frame a charter of robot rights? (This may seem a specious question when even human rights are so poorly protected throughout the world.) What happens when humanoid artefacts develop emotions, the impulse to laugh and to fall in love? Can a mental life only be sustained by the presence of a soul? Capek envisaged the possibility of a conflict between artificial systems and the human race (themselves artificially fabricated by gods.) In this he is reflecting the technological tensions that have always been seen to characterize Western industrial society; but other writers have sometimes chosen a different emphasis.

Asimov, well disposed to robots as he is, has worked to promote a positive image: in his stories robots can be friends, lovers,

* 1 A robot may not injure a human being or, through inaction, allow a human being to come to harm.
2 A robot must obey the orders given it by human beings except where such orders would conflict with the First Law.
3 A robot must protect its own existence as long as such protection does not conflict with the First or Second Law.

advisers (thus Gloria, in the tale *Robbie*, is distraught at having lost her pet robot: 'He was a *person* just like you and me and he was my *friend*. I want him back. Oh, Mamma, I want him back'). None the less the evident superiority of fictional robots to human beings may bring problems. In the short story *Reason*, the robot Cutie remarks that people are 'makeshift' in comparison: human beings are soft and flabby, apt to pass into a coma, and unable to withstand significant variations in environmental conditions; whereas Cutie works with 'almost one hundred per cent efficiency' ('I am composed of strong metal, am continuously conscious, and can stand extremes of environment easily').

It was inevitable that the arrival of film techniques would encourage the depiction of robots in the context of fiction and fantasy. Today there are hundreds of feature films that focus on the robot theme (Hefley and Zimmerman's book *Robots*, published in 1980, profiles dozens of such films and many more have been produced since that time). All possible variations are included: film robots can be seductive lovers, implacable warriors, wise advisers, confused servants and intrepid explorers. In this realm nothing is impossible, nothing ruled out: the mythmakers of the ancient world would be astounded.

By 1920 dozens of robot films had been produced, and even in such early days it was clear that many of the historical themes were being researched. *The Clown and the Automaton* was made by the legendary French film pioneer Georges Méliès in 1897; three years later he produced *Coppelia, the Animated Doll*, based on the Delibes ballet – and also on the ancient myth of Pygmalion. Soon the film robots were fulfilling their menacing potential. In two British films – *The Motor Valet* (1906) and *The Doll's Revenge* (1907) – animated artefacts variously attack furniture and human beings. And so on and so forth. The Coppelia theme is revisited time and time again, as is the familiar notion of humanoid machines that persistently run out of control. And early film-makers' knowledge of Jewish legend is aptly displayed by film titles (the golem appears in a number of pre-1920 German films and at least one from Denmark).

The absence of a soul in film robots appears to cause many difficulties (what would the early film-makers have done without the religious tradition?). For example, Georges Méliès's work contains the perennial confusion: in his film *Homunkulus* (1916) the robot is distraught, and driven to tyranny, when it realizes it does not possess a soul. Fortunately the world is saved from extinction by a convenient flash of lightning. In *Alrauna* (1918), another Méliès film, a scientist – no doubt inspired by Mary Shelley – decides to use artificial insemination to produce the daughter of a hanged criminal and a prostitute.

The 1920s saw more golem films, revolting robots, thinking machines, and at least one bright humanoid specimen that used its chess-playing skills to outwit a villain. Fritz Lang's *Metropolis* (1926) was one of the first films to carry a strong political message; in particular, it aimed to publicize the appalling conditions of industrial workers. The scientist-magician Rotwang makes a robot replica of Maria, the virtuous mediator in the underground city where the workers abjectly toil to service the luxurious metropolis above (Rotwang, graphically broadcasting a familiar robot theme, declares: 'I have created a machine in the image of man. . . . Now we have no further use for living workers'). The purpose is to counter Maria's efforts on behalf of the exploited workers. The robot is sent below to urge the workers to demolish the dikes, an act which destroys their homes and kills many of their children. In the chaos that follows, much of the city is ruined as the riot spreads to the surface. Rotwang is killed, the real Maria is freed, and the scene is optimistically set for better times.

In the 1930s there were further Frankenstein films, a golem sequel (a joint Czech–French production) and a number of productions that included the word 'robot' in the title. In one extraordinarily sexist film (*The Tin Man*, MGM, 1935), a robot is ordered to destroy all women. Fortunately it becomes confused and resolutely attacks its mysogynist creator. A 1940 production (*The Mysterious Dr Satan*, Republic) considers how an army of robots might be used to conquer the world, and further mighty humanoid specimens are wheeled out in such fifties' classics as *The Day the Earth Stood Still* and *Forbidden Planet*. The 1960s saw a spate of robots, devil dolls and at least one creature based on Greek mythology. *Jason and the Argonauts* (Columbia, 1963) offers an ambitious depiction of the bronze Talus, the giant devised by Hephaestus to guard Crete. Other engaging films introduce yet more robotic specimens, cyborgs, and the increasingly popular daleks, stars of the *Dr Who* TV series. A seminal film, *2001: A Space Odyssey*, was launched by MGM in 1968; and chess-playing robots, capable also of surgery on human beings, were encountered in the suitably ecological production of *Silent Running* (Universal, 1972). *Sleeper* (United Artists, 1973) had Woody Allen mixing with domestic robots of the future; in *Dark Star* (Bryanston, 1974) robot bombs are capable of philosophical discourse; and in the same year suggestions were made in *Westworld* (MGM) as to how robots of both sexes might be employed to serve as ideal lovers.

It is obvious that robots have been drafted to serve many roles in modern films, though the computer-based systems may in fact be realized by human actors (this is the cheapest ploy), by actual robotic devices (using wiring systems supplied by mainstream robot manufacturers) or by skilful special effects that may never have been within spitting distance of a real-world robot. Robots can sometimes be endearing pets (like R2D2 in the 1977 *Star Wars*), prodigiously talented servants (like Robby in *Forbidden Planet*), omnipotent warriors (Gort, able to destroy the world, in *The Day the Earth Stood Still*), pliant lovers (*Westworld* and *Blade Runner*), rapists (*Demon Seed*), spaceship crew members (*Alien* and *Aliens*), violent psychopaths (*Saturn 3*) and 'living' companions (*Short Circuit* and *Short Circuit 2*). There are few human social roles that have not been adopted and interpreted via the myth of robot culture.

Film robots are not only anthropomorphic in their structure, they also have essentially humanoid mentalities. The intelligent machines can envisage ends and goals; they can act with purpose. Some are quick to anger, others are capable of tender emotions and able to fall in love. In fantasy at least, there is already a wide profusion of artificial human beings. Moreover, the fiction and film robots have their counterparts in other art forms.

A 1924 ballet (*Machine of 3000* by Fortunato Depero) deals with the love of two locomotives for a stationmaster (humanoid emotions in a non-humanoid shell). The 'locomotives' are dressed in tubular costumes and required to move in a mechanical fashion. Eventually, their love unrequited (how could it be otherwise?), they are sent off in opposite directions. Sometimes there is a surreal merging of human and machine components, perhaps signalling the increasing humanlike attributes of artificial systems. Thus in the collages of Eduardo Paolozzi machine parts are assembled with human shapes; and, conversely, human figures are set inside mechanical contraptions. In such a way, mechanisms and people are induced to merge so as to form new entities with personality and intelligence. Again, efforts of this kind may be interpreted as an essay on the essential sameness of man and machine.

Robots often function as modern myths in children's comics, sometimes protecting the peace and sometimes attacking innocent folk (have they never heard of Asimov's First Law? – '*A robot may not injure a human being, or, through inaction, allow a human being to come to harm'*). In fact comic robots are part of a world-wide culture that already has a substantial history. For example, in 1934 the Japanese artist Masaki Sakamoto created the robot Tanku Tankuro, who was to feature widely in children's comics. His body, supporting a suitably humanoid head, resembled a bowling ball with holes out of which guns and swords could be produced as required. Tanku was always on the side of Good – which means that in the turbulent Asia of the 1930s he was always keen to support The Japanese Way. When Japan found that its imperial ambitions were being thwarted in the 1940s, robots were sometimes created in comics and magazines as fantasy warriors – perhaps a sublime form of wish fulfilment! Thus in 1943 the artist Ryuichi Yokoyama produced a propaganda cartoon called *kagaku senshi Nyu Yoku ni shutsugen su* (*The Science Warrior Appears in New York*) in apparent frustration at Japan's inability to counter US air raids. The science warrior, bestriding the skyscrapers, is a fearsome fellow with massive spikes on his metallic boots; he is able to fire simultaneously from numerous apertures and protuberances, and New York is shown crumbling under the onslaught.

A more peaceful robot, Tetsuwan Atomu (Mighty Atom), was created by a young artist/medical student in 1951. Osamu Tezuka's humanoid specimen proved so popular that it appeared for a run of eighteen years in a comic book story, starred in an animated television series in 1963 and was later exported as Astro Boy. Mighty Atom engagingly explored the possibility of a machine-human social symbiosis: he was a robot child in a 'normal' (robot) family, but went to school with human children. Like the science warrior he was well equipped – searchlight eyes, rockets in his feet, and a rear machine gun – but only waged war on unambiguous villains such as bandits and monsters. Here was a humanoid machine with feelings, a powerful ally devoted to justice. Perhaps Isaac Asimov would approve.

However, even the obvious benevolence of Mighty Atom could not disguise the possibility that powerful robots still needed to be regulated: there was as much awareness of the Frankenstein problem in the East as there was in the West. In consequence Osamu Tezuba advanced his Ten Principles of Robot Law (aficionados might like to compare these with Asimov's Three Laws):

1 Robots are created to serve mankind.
2 Robots shall never injure or kill humans.

3 Robots shall call the human that creates them 'father'.
4 Robots can make anything, except money.
5 Robots shall never go abroad without permission.
6 Male and female robots shall never change roles.
7 Robots shall never change their appearance or assume another identity without permission.
8 Robots created as adults shall never act as children.
9 Robots shall never assemble other robots that have been scrapped by humans.
10 Robots shall never damage human homes or tools.

In 1956 another Japanese artist, Mitsuteru Yokoyama, created Tetsujin 28go (or Iron Man No. 28), a giant metal monster controlled by a device operated by a boy 'private detective', Shotaro Kinta. This robot had no artificial intelligence, no element of autonomy: the iron man was only a machine – a departure from many of the other fanciful humanoid systems.

It is obvious that the mythical and fictional robots are largely independent of the current state of practical technology: in fantasy there is no need to remember the laws of physics and chemistry or to acknowledge the limitations of current manufacturing technology. But at the same time the robot myths and fictions are significantly mediated by the current state of science and technology. In the ancient world it was enough to invent mythical humanoids that were generated and animated by magic: supernatural and occult powers were all-purpose devices that could be drafted to fill the technological gaps (in fact they filled nothing, explained nothing). By contrast, in the modern age robot fantasy draws heavily on what technology has shown to be possible in the real world. In today's film and fiction we can manufacture humanoid systems on assembly lines, much as this is accomplished in the factories of Europe, the United States and Japan. And the fantasy, as always, stimulates the human imagination, keeping alive the dreams of what artificial systems may become in the years ahead.

There is a two-track approach to the development of truly humanoid artefacts: on the one hand lies the nebulous realm of fantasy and fiction, allowing all the extravagances of imagination and dreams; on the other is the world of practical invention, the creation of physical automata, real-world robots. As with fantasy, there is a substantial tradition of practical automata, skilfully contrived inventions that can simulate animation (such devices are considered in Chapter 2). The true ancestors of the modern robot are to be found in historical automata.

It is not always realized that such concepts as robotics and artificial intelligence are not newly hatched in the modern age but extend back over the millennia. In this chapter examples have been given of ancient writings that prefigure the practical intelligent robots that are being designed and manufactured throughout the world. The ancient Sanskrit *Mahabharata* features a female robot of surpassing charms, and in another Asian work, the *Brihatathasaritsagara* of the eleventh century, there are mechanical humanoids that can speak and dance. And Hephaestus, smith to the Greek gods, has also been encountered, who manufactured golden maidens who could speak and walk, autonomous tripods that could run to serve the gods, and the bronze Talus, mighty guardian of Crete. Hephaestus was not only a robot engineer: in supplying the gods with their accoutrements he also manufactured chariots, sceptres, armour, arrows, swords, necklaces, goblets, vases, bronze bulls whose nostrils spurted flame, and gold and silver dogs to guard the celestial palaces.

It is also significant that the ancient myths, from many different cultures, try to deal with the matter of divine creativity. Many legends tell of an ancient world of gods, often exercised by their own conflicts and tribulations, where suddenly the first human beings are envisaged, created and set upon the earth. It is easy therefore to depict men and women as artefacts, fabricated by the celestial engineers out of clay, earth, rock, wood, excrement, gouts of blood or pieces of bone. The ancient gods were the first robot engineers, keen to invent replicas of themselves – warriors, lovers, travellers – to function in divinely conceived games, to further the godly aims and purposes. And then there was the perennial problem of spirit and soul, the mysterious elements that consigned intelligence, animation and sweet immortality.

For the gods it was a simple matter: they merely had to breathe upon an inanimate form and life would be bestowed. Or perhaps a god would laugh, or play a drum, or utter a special proclamation. In the ancient myths there were many ways in which the magic wand could be induced to work, but in the earliest days it was clear that the magical tricks could only be managed by divinities, by those supernatural creatures for whom most (if not all) things were possible.

However, human arrogance developed through civilization. The seeds were evident in the most primitive societies where priests, often sanctioned by nothing more than nepotism or dynastic claim, had unique knowledge of divine wishes – they alone knew how to tap into cosmic power. And the early arrogance of priests was later matched by the burgeoning confidence, however misplaced, of necromancers and occultists. Perhaps the secret of life was not only a matter of divine discretion. Perhaps there was a route to magic quite independent of godly behest and ritual pieties. So the occultists searched for methods and principles that *they* could understand and control, without the sanction of priestly authority. The dream was an age-old one – how first to shape a formless mass, and then to animate it; in other words, how to create life.

There seemed to be many possible routes to the realization of the dream. Perhaps the gods had shown the way: they, after all, knew about creation. So what did the sacred writings say? How were the words used? Were there cabbalistic clues to how mere symbols, framing mystical words, could achieve desired goals? In an obvious sense the cabbalists and occultists were striving to develop naturalistic techniques. Their biochemical brews often relied upon the supposed vitalities of blood, urine, seminal discharges or menstrual fluids. Their biochemical procedures far surpassed in complexity anything that Prometheus or Jehovah had attempted – so perhaps the occultists were hinting at one of several *scientific* routes to the creation of artificial life.

The myths also suggested, with massively increasing confidence in the modern age, that there were discoverable practical ways in which mere mortals could set about creating artificial men and women: modern film and fiction, as has been seen, are full of such suggestions. In fact the modern age is unique in its capacity to frame artefacts capable of humanlike behaviour, not as simple mimicry but in ways that signal true intelligence. After all, the ancient gods and their modern counterparts are fictions, unreal fabrications designed to meet real needs. To find the true robot pioneers it is necessary to consider the practical human engineers, those skilful inventors prepared to dispense with mystery and magic and to accept the mundane constraints of natural law: It is time to consider the makers of automata.

2 Early Automata

The concept of the working automaton, like the mythical artificial humanoid, goes back through the centuries. It can be found in almost every sort of early society, even in those widely separated by custom and historical experience. The type of automaton, of course, varies from one place to another, and semantics soon rears its head – just what is an automaton? And in what way can it be regarded as an ancestor of the modern robot? Before looking at particular words let's glance at the nineteenth century, a relatively late period in the history of automata.

In *The Uncommercial Traveller*, written in the 1860s, Charles Dickens used the word 'automaton' where today he might have said 'robot'. While exploring London churches he encountered a young child who sat motionless throughout a service ('closely fitted into the corner, like a rain-water pipe'). The child was accompanied by a man who apparently discouraged youthful activity. Confronted by the aspect of the motionless child who never smiled, Dickens commented: 'Once, the idea occurred to me that it was an automaton, and that the personage had made it . . .', but once the child had spoken, Dickens abandoned the notion. So a true automaton cannot indulge in discourse with a human being. Is this true? We shall see. Another Victorian writer, Samuel Butler, mentioned numerical automata in *Erewhon*, first published in 1872: 'Our sum-engines never drop a figure . . . the machine is brisk and active . . . clear-headed and collected . . . it needs no slumber . . . ever at its post, ever ready for work, its alacrity never flags, its patience never gives in; its might is stronger than combined hundreds . . .'

The word 'automaton' derives from the Greek *automotos* (literally, self-moving). In a 1671 English dictionary 'automatous' was defined as having a motion within itself. The idea is simple enough. It is easy to envisage how a machine, once set up, can function on its own. A mechanical clock, for instance, equipped with a spring, is a type of automaton; in fact the discoveries made in early clockmaking were very important to many families of automata that had an entirely different purpose. Machines could be designed with springs or weights, or could carry a reservoir replenished by rain, to facilitate independent movement. And indeed it has been in particular the movement of mechanical and then electronic automata over the ages that has proved so evocative to human observers. Autonomous movement creates the illusion of animation, which is why so many early automata were made in the form of birds, dolls or other creatures. In modern times movement has often been exploited to create the impression of life; as, for example, with *The Senster*, a cybernetic sculpture constructed by Edward Ihnatowicz in 1970 for the Philips Evoluon in Eindhoven in the Netherlands. The device – an automaton relying on mechanical, electrical and electronic effects – was shaped in the form of a lobster's claw, consisting of six hinged joints activated by electrohydraulics to allow wide freedom of movement. It was found that the sudden unpredictable motions of the structure encouraged observers to think that it was alive. Thus the cybernetician Jasia Reichardt (1978) commented: 'It is as if behaviour were more important than appearance in making us feel that something is alive.' This is a characteristic feature of many classes of automata over the centuries.

Modern definitions of 'automaton' preserve the idea of a self-moving machine, a device that contains within itself mechanisms and structures that can yield motion in appropriate circumstances. The motion may be largely predictable, as with most clockwork devices, or it may be possible to incorporate an element of spontaneity and unpredictability. An automaton may, for example, be able to respond to environmental conditions, as indeed *The Senster* monitored the presence of people in its surroundings and behaved accordingly. The modern electronic computer has given automata an immensely enlarged scope for seemingly spontaneous behaviour in the real world.

Historical automata invariably incorporated state-of-the-art technology, produced by skilled engineers with clear insights into scientific principles. The ancient Greeks, for instance, were well acquainted with gears and levers, and well equipped to exploit the principles of pneumatics and hydraulics. Theoretical treaties were produced, explaining the principles and carrying illustrations of working automata. But usually the practical devices were engaged in trivial pursuits, designed merely to amuse the onlooker or to embellish the homes of the wealthy. Seemingly autonomous birds, manufactured out of wood or metal, could be induced to sing at specified times, artificial figures could be activated in minuscule theatres of automata, and humanoid creatures, suitably decorated, could be organized to track the hours of the day.

The culture of automata was a worldwide phenomenon, as indeed it is today. The ancient Greeks and Romans were well acquainted with the design and fabrication of self-moving machines, as were the Arabs, the Chinese, the Japanese and the Indians. There is a common technology of automata in the ancient world, though the paths often diverged in later centuries. At first, machines could be designed to move by the simple pressure of weights, wind and water; then gears and levers were skilfully juxtaposed to give greater flexibility and variation; and when the spring was invented in the third century BC – a useful source of stored energy – the scope for automata was massively increased. In Switzerland, for example, centuries later, mechanical timepieces were constructed as animate humanoids to attract rich patrons. New families of automata, nicely exploiting mechanical and hydraulic principles, appeared in the gardens and houses of the wealthy.

Some of the devices were extraordinarily ambitious. There were humanoid automata that could play music on keyboards and the violin, ducks that could 'digest' food, self-moving machines built to resemble angels (these latter devices, designed by the inventor Jaques de Vaucanson, caused a religious scandal, just as the artificial humanoids of the Spanish Jaquet Droz led him to be accused of witchcraft). With the dawn of the industrial revolution and the spread of science and technology, the self-moving machine came to acquire a more productive significance. Today the simple mechanical automaton is largely confined to history and to toyshops; but electronic automata – the digital computers of the modern age – are essential to the quality of life in the developed world.

Origins

The first humanoid representations to be created were undoubtedly the simple sculpted figures and wall paintings in prehistoric society. Such representations had no capacity for autonomous movement, though early men and women, confused by drug-induced ritual, may have thought differently. Somehow the sculpted or fabricated figures had to evolve internal mechanism that would allow autonomous motion in the real world, not in the world of fantasy, delirium or dreams. First a few mechanical principles had to be discovered and exploited, a tortuously protracted development that says something about the limitations on human creative potential.

Small figurines were provided with movable limbs as early as four millennia ago: examples have been discovered in Egyptian tombs dating to around 2000 BC. Simple levers could be operated to agitate the limbs of artificial tradesmen, so that the wealthy dead could continue to be well served in the afterlife. The Greek philosopher Plato described puppets with movable legs and arms, and some could also move their eyes and mouths. And masks with articulated jaws, used by priests and others in religious ritual, are associated with many early cultures. The old Greek traveller Pausanias, writing in the second century AD, described a mechanical eagle made of brass which was said to rise aloft when a mechanism was activated. Over a period of centuries the simple levers and articulations of primitive societies were able to evolve into the complex automata of more sophisticated communities.

The classical world produced a host of scientists and engineers – including such figures as Euclid, Archimedes, Ctesibus, Philon and Hero – many of whom were interested in the design and manufacture of automata. Importantly enough, they were also concerned to develop the underlying scientific principles, helping to found in this context many physical sciences. More than two thousand years ago Hero of Alexandria wrote his seminal *Treatise on Pneumatics*, a prodigious work in which he describes a wealth of automata such as moving figures and singing birds – a world that one modern writer on robotics has dubbed 'an Ancient Greek Disneyland' (Scott, 1984). It is known that Hero used physical models to illustrate scientific principles, and it is likely that some of the models were working automata (the nineteenth-century Frenchman Poyot managed to build a number of models first designed by Hero). He described animated birds, beasts and people, sometimes as solitary artefacts and sometimes operating in groups. For example, his treatise gives attention to a mechanical theatre in which the figures are worked by weights and pulleys, exploiting principles that were to be used extensively by later European clockmakers. And elsewhere Hero described theatrical scenery that can be changed using a variety of mechanical systems. In Pindar's *Olympic Ode*, *c.* 520 BC there are further descriptions of animated figures that seemingly adorned every public street and which appeared 'to breathe in stone, or move their marble feet'. And the Athenian Daedalus, an apparently real historical figure but with many mythological connections, is said to have invented the axe, the wedge, the level and many other mechanical devices; he is also said to have created a bronze warrior and a wooden humanoid device animated by a flow of mercury through concealed channels (great wooden statues carried in processions were called *Daedala* as a tribute to the inventor). Perhaps such claims are more plausible than another celebrated tale connected with the name of Daedalus – that he made wings for himself and his son Icarus and attached them with wax, that he flew successfully across the Aegean Sea, but that Icarus flew too close to the sun, melting his wax and causing him to crash to earth.

Seemingly animate statues have also been used by priests to maintain control over gullible people. The use of ancient statues (dating to around 2000 BC) containing trumpets, through which the priests would address their ignorant audiences, can be compared with the modern exploitation of 'weeping Madonnas' to impress superstitious believers. In the ancient world, as today, not everyone was convinced. Celsus, for example, one of the great sceptics of those times, was keen to decry magic and animals 'not really living but having the appearance of life'. When the engineers gradually took over from the magicians and priests a fresh spate of seemingly animate inventions, relying on obvious physical principles, was still able to generate superstitious awe.

In ancient China, automata were described that were akin to those that would emerge in other lands. Thus King Shu Tse, around 500 BC, described such wooden automata as a flying magpie and a horse operated by springs. At about the same time (*c.* 400 BC) in the West, Archytas of Tarentum, who is said to have invented the screw and the pulley, created a wooden pigeon suspended from the end of a pivot that was rotated by a jet of water or steam, so simulating flight. In the fourth century AD a golden Buddhist statue, accompanied by animated models of Taoist monks, was carried through the streets to impress the pious. And other Chinese records tell of a mechanical monk able to extend its hands, cry 'Alms! Alms!', and place coins in a bag when they reached a certain quantity.

It was claimed that a wooden otter, built in China around AD

790, could catch fish; just as a wooden cat (*c.* 890) was supposed to be able to catch rats and dancing tiger-flies. One engaging tale concerns the prudent inventiveness of a certain Prince Kaya, son of the Japanese Emperor Kanmu. He is said to have made a mechanical doll with a large bowl and set it in his rice paddy. When the rain had filled the bowl, the doll would lift it up and pour water over its own face. This so fascinated the simple people of Kyoto that they kept filling up the bowl to watch the behaviour of the doll. In this way Prince Kaya ensured that his rice paddy was frequently watered.

Descriptions of machines and the associated mechanical principles can also be found in the Indian tradition and from Arab lands (daunted by Muslim turmoil in the modern world, we often forget the rich scientific legacy of Islamic culture). Thus the Indian Prince Bhoja described in the eleventh-century *Samarangana-sutradhara* many details of machine architecture, including devices constructed to resemble animals and human beings. In a contemporary text an effort is made to relate the machines to what were presumed to be the five cosmic elements: 'The yantras [machines] based on earth materials undertake activities like shutting doors; a water-based yantra will be as lively as living organisms; a fire yantra emits flames; an air yantra moves to and fro; the elements of either serve to convey the sound generated by these yantras.'

The water clock (*clepsydra*) was widely used in ancient Egypt and amongst the Greeks and Romans. The operating principle was simple enough: water in a graduated tube was slowly allowed to escape from the bottom of the vessel, so indicating the passage of time. More complicated *clepsydras* included a float on the surface of the receding water, providing an effective pointer to a figured dial – the ancestor of the modern clock face. There was obvious scope here for the development of water-driven automata, and as early as 500 BC the Egyptians had constructed ingenious water clocks in the form of urinating apes. Water had long been used as an activating element, but often for merely ornamental devices; by contrast the *clepsydra* had immense practical value. (A *clepsydra* presented by the King of Persia to the Emperor Charlemagne in AD 807 was activated by little balls that fell on to a brass drum. At particular times twelve miniature horsemen were caused to march round the dial.)

In 835 a throne built for the Byzantine Emperor Theophilus was decorated with lions that roared and birds that sang; just as in 827 a golden tree carrying singing metal birds was built for the Caliph Abd-Allah-Al-Mamoun. It is clear that such devices were amusing and diverting, rather than intended to serve any useful function. Many such inventions were described in the *Book of Knowledge of Ingenious Mechanical Devices* by the Mesopotamian Ibn Al-Razzaz Al-Razzaz, writing in the thirteenth century and drawing on material compiled by Al-Jazari by command of the Sultan Mahmud at Amid on the Upper Tigris. The *Book of Knowledge* describes *clepsydras,* fountains and a wide range of other hydraulic devices. Another Muslim invention was the 'Peacock Fountain', an ingenious device for washing the hands. Here the body of the peacock serves as a reservoir filled via a hole in the tip of its tall. When the bird's beak is pulled, water pours into a basin and from there to a cistern beneath, Carefully sited floats control the whole operation and the behaviour of two figures: one emerges with a bowl of perfumed powder, after which the second comes out to offer a hand towel. Other models attributed to Al-Jazari included a device for dispensing wine. As always, the patronage of the wealthy stimulated the design and construction of mechanical automata.

Throughout the ancient world there was, perhaps unexpectedly, a shared culture of automata, which at best represents a burgeoning framework of technology that was able to evolve to generate the sophisticated artefacts of the modern age. Perhaps the shared culture arose through independent initiatives in different parts of the world, or perhaps a few seminal ideas were carried abroad through trade and other forms of contact. In any event there are many sites of historical automata, and it is worth remembering that the early engineers – Chinese, Japanese, Greek, Egyptian, Ethiopian, Babylonian, Indian, Arab and others – came to contribute to the emergence of a ubiquitous world culture. The further examples given in this chapter are necessarily few and highly selective: many other examples could be chosen to illustrate what is an immensely rich international tradition.

Further Asian Ingenuity

Whether or not the earliest automata originated independently in different parts of the world, it is clearly true that in the centuries that followed there was considerable cross-fertilization of ideas between one region and another. For example, European exploration and expansion introduced aspects of technology to Chinese and Japanese cultures that were already highly sophisticated in many ways. It has been suggested (by the writer and translator Frederik Schodt, 1988, for instance) that when the first Portuguese adventurers landed on the Japanese island of Tanegashima in 1543 they encountered a technology which in many respects was superior to that of the country they had left behind. For instance, the Japanese were particularly skilled in ceremics and swordmaking, but had much to learn from the West in the manufacture of guns and clocks. It is recorded that the Spanish Jesuit missionary St Francis Xavier introduced the first mechanical timeplace into Japan in 1551, to bribe a local lord into allowing him to open a mission. The Japanese had already obtained knowledge of levers, pulleys and springs from China, but with the arrival of the Europeans they were confronted for the first time with complex and intricate automatic mechanisms, wound springs and the escapement governor. And it was inevitable that many Japanese inventors were to learn from clockmaking how to design and construct automata.

As early as 1662 the craftsman Omi Takeda began using automata in the entertainment district of Osaka: he is said to have devised a tea-carrying doll and to have organized bawdy displays. The Takeda *karakuri* ('automata') became widely known, his family continuing the tradition for nearly a century. The displays of automata included mechanical carp that swam and then jumped out of the water; and intricate dolls that could shoot darts from blowpipes, hold brushes in each hand and their mouth to write characters simultaneously, climb ladders and even urinate on stage. In devising such automata it was always useful to try to learn from nature, a principle that is followed by many robotics engineers today. Thus the author Yorinao Hosokawa, influenced by Chinese ideas, recommends in his *Karakurizui* (1796) that inventors observe the world around them: 'The rudder, for example, was made by observing the action of a fish tail; oars were created by observing the sideways motion of its fins. Zhuge Liang [a second-century Chinese politician] observed the doll that his wife made and created an automatic ox-cart. Omi Takeda observed children playing with sand and used the idea to power his mechanisms.' And just as Western timepieces had been transported to Japan, to influence the development of automata, so they were conveyed to China.

The Jesuit Father Matthew Ricci, head of a Chinese mission in

1601, wanted to introduce Western clocks to the Chinese court to provide evidence of the value of Christian civilization; however, the court eunuchs were worried that they would not be able to manage the complicated machinery. Later Jesuits, pandering to the Chinese fascination with automata, manufactured a range of mechanical toys for the court. It is recorded, for example, that Father Gabriel de Magalhaens built a mechanical walking man for Emperor K'ang Hsi which, carrying a sword in one hand and a shield in the other, was operated by internal springs. Similarly Father Jean Mathieu de Ventavon, who reached China in 1766 during the reign of Ch'ien Lung, constructed various clocks and automata, not least two walking men able to carry a vase of flowers ('the work of not less than two years . . . with which the Emperor was very pleased').

Such devices, and the complicated mechanical gifts from European sources, accumulated in the Chinese court, entertaining the wealthy and stimulating the inventiveness of the court craftsmen. A clock presented to Ch'ien Lung by the East India Company included chariots that carried seated figures. A bird, encrusted with rubies and diamonds and perched on a lady's finger, fluttered its wings when a diamond button was pressed; and, when hidden machinery was activated, the chariot could be moved in any direction, apparently being pushed by a boy behind it. Alas, when in 1860 a force of French and British troops stormed the Summer Palace many artefacts, including priceless automata, were plundered and destroyed. A contemporary observer, the Comte d'Hérisson, recorded in his diary: 'Every trooper had his bird, his music box, his alarm clock and his rabbit . . . everywhere bells were ringing.' One looted treasure was an intricate music box given by Marie Antoinette to the Chinese Emperor.

In another military confrontation, the capture of Seringapatam in India in 1799, the famous automaton known as Tippoo's Tiger fell into the hands of the British. This device, now in London's Victoria and Albert Museum, is a model of a wooden tiger mauling a prostrate European. It carries an organ played by a mechanical keyboard and operated by bellows and a reservoir of air. The tiger growls and the man screams, appropriately enough, when a handle is turned. The original owner, the ruler of Mysore, was known to have had dealings with the French, and there is some suggestion that the mechanism may be of French origin. Such devices, whether of indigenous or foreign origin, formed part of a growing family of mechanical contrivances – in Japan, China, India and elsewhere – that in some cases can be shown to be the direct ancestors of complex automata in the modern world. It is noteworthy that one of the *karakuri* masters established a firm that would one day emerge as the modern corporate giant Toshiba. . . .

When the British interpreter Ernest Satow visited Nagasaki in 1867 he recorded in his diary a meeting with a certain Hisashige Tanaka: 'Originally a Kioto clockmaker, who had developed into a skilled mechanical engineer, and had constructed engines and boilers for a couple of Japanese steamers.' In fact Tanaka was one of the great automata makers, a celebrated *karakuri* master. In mastering 'basic Western mechanical principles through sheer effort . . . he symbolically created a technical bridge between feudal and modern Japan' (Schodt, 1988). Following Omi Takeda he took *karakuri* shows around Japan, at one time even experimenting with steam to power them.

The significance of Tanaka was that he went far beyond the creation of simple automata: other *karakuri* masters, often highly skilled craftsmen, advanced few technological principles and had no enduring influence. By contrast, Tanaka was an inventor of genius, able to develop a range of artefacts that had real technological significance. For example, he used technology from an improvised air gun to develop an 'endless' oil lamp that relied upon compressed air to keep vegetable oil burning as it travelled up the wick. He also invented a fire extinguisher, a miniature planetarium (embodying in 1847 the contemporary Buddhist theory that the sun revolved round the earth), and the Mannendokei, an 'eternal clock' that could be wound to run for 225 days (the clock carried six faces giving Japanese time, Western time, the phase of the moon, the day, the date and other details; and inside a glass dome the sun and the moon revolved over a map of Japan).

Tanaka was sufficiently astute to perceive the current limitations of Japanese technology: details of Western technological progress were percolating into Asia, and he obtained what information he could from Western sources. Dutch texts were of particular value. Japanese scholars translated Dutch and other Western works and in this way a growing body of scientific and technological knowledge was made available. In 1852, by studying a Dutch reference book, Tanaka managed to build a model steamship. He later helped to construct a paddle-wheel steamer, other types of ships and military equipment.

Towards the end of his life Tanaka formed the Tanaka Seizojo company to manufacture and repair telegraph equipment (a sign boldly declared: 'Orders Taken for Design of Any and All Machines'). The firm eventually developed into Toshiba, one of the mightiest modern corporations and one of the largest makers of industrial robots. Today the company's debt to the past is nicely acknowledged by the replica of a seventeenth-century automaton, a tea-carrying doll, displayed in a glass case in the Toshiba Manufacturing Engineering Laboratory. Toshiba now uses industrial robots to manufacture a wide range of products – televisions, video cassette recorders, electronic blackboards and so on: the early *karakuri* masters – Omi Takeda, Benkichi Ono, Hisashige Tanaka and the rest – would be amazed.

Towards Mechanical Life

Following the development of early automata in the ancient world, great strides were made in Europe and elsewhere. It was said, for example, that the philosopher Albertus Magnus (1204–72) had built a functional robot servant, and there were several versions of the tale. In one account his pupil St Thomas Aquinas, meeting the robot on the street and thinking it to be the work of the devil, smashed it to pieces (this seems to have been the common response of pious observers: a nervous sea-captain is reported to have thrown overboard an 'animate' automaton, *'ma fille Francine'*, constructed by the philosopher René Descartes around 1640). Roger Bacon (1214–94) is said to have constructed a talking head.

Sometimes, in contrast to the experience of Albertus Magnus and Descartes, pious believers were able to find a constructive use for simple automata. Medieval miracle plays were often acted out with mechanical puppets, and in some churches the preacher was able to use foot pedals to activate mechanical models of Jesus and the Madonna (a wooden Christ that moved its head, shifted an eye and lolled out a tongue would stand a fair chance of terrorizing a superstitious congregation). And many of the devices could exploit clockwork to achieve the desired effects. For example, clock Jacks ('Jaquemarts') were widespread on churches and elsewhere from the fourteenth century. A soldier in armour, mechanically activated, was placed on Southwold church in Suffolk around 1480, and a number of automaton figures were placed on St Paul's Cathedral and elsewhere (there is a reference in Shakespeare's *Richard II* to 'Jack of the Clock').

Some of the Jacks were immensely ambitious, designed with

high imagination and constructed with loving care. Thus a Berne city clock, dating to 1530, carries a parade of bears that circle around a turntable each hour; a cockerel crows three times and flaps its wings. Many of the early clocks were worked by heavy lead weights that had to be manually raised each day. It was essential, of course, that such devices should remain within the conceptual grasp of the church clerics: any invention that was too advanced might be deemed spiritually suspect (we feel with Albertus Magnus, his painstaking invention destroyed by Aquinas, when he cries: 'Thus perishes the work of thirty years'). When Leonardo da Vinci tried to realize Roger Bacon's dream of a 'bird-man', there were no doubt pious protestations that the creation of living things was a divine prerogative. When Gianello della Torre of Cremona, a skilled mathematician and scientist, created artificial sparrows that could fly around the room, he was accused by a Mother Superior of having occult powers. His only extant work, held today in the Vienna Kunsthistorisches Museum, is the automaton of a lady lute player, an elegant wooden figure with an iron mechanism hidden beneath her skirts. This clockwork figure can be set to walk in a circle or a straight line, its feet lifting one by one and its left hand plucking the lute. In the same spirit the famous inventor Hans Bullmann, who served Ferdinand I, the Holy Roman Emperor, in the sixteenth century, created humanoid automata able to beat tambourines and play the lute.

A proliferation of automata were designed to resemble animals and human beings. A celebrated duck, devised by Jacques de Vaucanson (1709–82), was made of gilded copper; it was claimed that this invention 'drinks, eats, quacks, splashes about on the water, and digests his food like a living duck'. In a single wing there were more than four hundred articulated pieces, and when the duck fell into disrepair it took more than four years to bring it back into full working order. From a contemporary report, cited by Chapuis and Droz (1958), we learn that: 'After each of the duck's performances there was an interval of a quarter of an hour to replace the food. A singer announced the duck. As soon as the audience saw it climbing on the stage, everyone cried: "Quack, quack, quack". Greatest amazement was caused when it drank three glasses of wine.' The same inventor produced a range of other devices, including a talented flute player able to play twelve different tunes by exploiting a current of air fed through the complex mechanism. Another inventor, Friedrich von Knauss (1724–89), created a wide range of automatic writing machines, talking machines and writing dolls. Of a talking machine built by the inventor Baron Wolfgang von Kempelen (1734–1804), the celebrated poet, German dramatist and amateur scientist Goethe remarked: 'The talking machine of Kempelen is not very loquacious but it pronounces certain childish words very nicely.'

A showman called Defrance gave automata performances at the Tuileries Palace in Paris in 1746. Life-size figures of a shepherd and a shepherdess, with moving lips, were able to play no fewer than thirty different tunes; and his birds were equipped to sing 'several airs with marvellous delicacy'. These automata were later acquired by the physician and sculptor Blaise Lagrelet, who demonstrated a number of 'electrical experiments'; later he gave exhibitions of electrically driven boats and electrically driven dancing puppets. One automaton, constructed to resemble Mercury, was able to respond to the audience by answering 'yes' or 'no' to questions. Similarly, in shows at Parisian fairs, the conjuror Toscani exhibited little figures that 'imitated all natural movements'; such displays included artificial sailors, able to swim for their lives. The promotional advertisements claimed that the figures were not worked by wires or 'other means', so there is still room for speculation as to how the intriguing results were achieved. A show staged in Paris in 1778 included a box that could lock itself, a lamp that could turn itself off, and an artificial huntsman capable of firing arrows. And at the St Germain fair of 1749 Jacques de Vaucanson displayed various automata, including an artificial woman able to raise a glass to receive wine poured from the bill of an artificial pigeon; a mechanical shopkeeper able to bring to the audience whatever they ordered; and a Moorish character able to sound a bell in one hand by swinging a hammer held in the other. Vaucanson later became the Inspector of Mechanical Inventions at the Académie Royale des Sciences, and was an acknowledged pioneer in the development of machine tools.

By the end of the eighteenth century many European craftsmen, the Western counterparts of the Asian *karakuri* masters, were working to produce mechanical automata that could imitate many types of human activity. Swiss craftsmen, for instance, were constructing automata that could write words, draw pictures and play various types of musical instruments. The inventors Pierre and Henri-Louis Droz were notable among such craftsmen. In 1770 they constructed their Scribe, a mechanical child that was able to dip a quill pen in ink and then write on a piece of paper. The artificial Draughtsman, built in 1773, using complex precision machinery, could draw various pictures, including a portrait of Louis XV; and the artificial Musician, equipped to play the organ, had articulated fingers, a heaving breast and eyes that could glance from side to side. By this time such devices were relying on sophisticated principles, often using delicate clockwork, precision cams, skilfully positioned levers and other mechanical elements to implement relatively complicated systems. Detailed plans were submitted to such bodies as the Royal Society in London and the Académie Royale des Sciences in Paris for the construction of clocks, mills, musical instruments, military equipment and calculating machines (these last would later be seen as the ancestors of modern electronic computers). In 1731 a Monsieur Maillard presented designs, in *Machines et Inventions approuveés par l'Académie des Sciences*, for chariots with the back wheels worked via cogs and handles; in 1733 he offered plans for an artificial swan, a seahorse, and a cart pulled by an artificial horse. Such efforts gave further impulse to a growing range of inventions that increasingly relied upon detailed scientific and technological principles. Goethe believed that his times were singularly blessed by scientific insight ('What wonderful luck to have lived through the second half of the eighteenth century').

In 1846 Euphonia, a talking machine in the form of a bearded Turk, was exhibited in the Egyptian Hall, Piccadilly, London; this humanoid specimen had been built over a twenty-five-year period by Professor Faber of Vienna. The device contained levers, keys, a double-bellows and a movable mouth with a flexible tongue and an india rubber palate; and it was able to produce humanlike sounds, laughing, whispering and singing, and asking and answering questions. It is interesting that, because Faber was a native German speaker, Euphonia spoke English with a pronounced German accent. Various other automata of the day were equipped to talk, to move arms and legs, and to play games (Kempelen had earlier manufactured a chess-playing automaton with cogs, levers, gears and moving cylinders; but had cheated somewhat by including space inside the device to accommodate a human operator!). Euphonia (sometimes referred to as 'The Euphonis'), however, remained one of the most impressive devices of the time. To achieve his effects, Faber used 'a caoutchouc [rubber] imitation of the larynx, tongue, nostrils and was able to operate it very skilfully even imitating song' (*Chambers Magazine*, 1945). Alas, there is no sense in which the device was truly autonomous: an operator, along with the controlling mechanism, was hidden behind a curtained framework.

The intricately constructed automata were characteristically activated by weights, springs, pressure of fluids and so on. They worked according to nicely controlled sequences of operations, the ordered procedures that would eventually evolve into the programs of the modern digital computer. By the end of the nineteenth century it was manifestly obvious that there were serious limitations to the degree of autonomy that could be built into purely mechanical systems. The principles of reasoning had already been enshrined in various symbolic systems (via the lengthy evolution of logic from Aristotle to such nineteenth-century thinkers as Boole and De Morgan), and now there was a pressing need for a new technology, for a body of methods and techniques that would allow truly impressive levels of autonomy to be built into artificial systems. The new technology was to be electronics, without which modern robots could never have emerged.

Early in the twentieth century an electromagnetic chess-playing automaton, able successfully to conclude a simple end-game, was constructed by Leonardo Torres y Quevado, President of the Academy of Sciences in Madrid; here, in a characteristic end-game, the white king and rook could mate the black king from any position. Quevado's celebrated machine, a triumph of classical mechanics, was able to defeat the great cybernetician Norbert Wiener at the 1951 Congress of Cybernetics in Paris. This remarkable event was seen as the last victory of classical mechanics over the rapidly developing, electronics-based, modern cybernetics. It was necessary for automata to pass through a lengthy phase of mechanical, electrical and electromagnetic evolution, but the truly autonomous machines of the modern age – 'thinking' computers, decision-making robots and so on – were only made possible by the peculiar features of electronics technology. But before discussing the early days of electronics, it is worth glancing at two other historical threads that are important ancestral influences in the emergence of the modern robot: the development of 'living dolls' and the evolution of mechanical computers, the ancestors of the modern robot brain.

Living Dolls

With the *karakuri* and other allusions, various types of 'living dolls' have already been encountered. More can be said, of some importance to the theme of this book: the doll is a nicely anthropomorphic artefact, at least a cousin of the intelligent robot. It is obviously true that most historical and modern dolls have no autonomous capability, but some are more competent than others, incorporating mechanisms that can create the illusion of life. The sophisticated doll, fabricated by ancient craftsmen or in modern factories, at least stimulates the idea that humanoid artefacts can be designed to behave in surprisingly human ways.

By the beginning of the nineteenth century techniques had been developed for the creation of a wide range of humanoid automata. For example Leonard Maezel, inventor of the metronome, constructed a range of life-size automata. In 1804 he devised an artificial trumpeter able to 'tongue' his instrument to play cavalry marches; and in 1805 he created a full musical orchestra (the Panharmonicon) with no fewer than forty-two players, including those for trumpet, flute, clarinet and triangle. It was inevitable that the fabrication techniques should feed into the construction of automated dolls, humanoid figures capable of being handled and manipulated for amusement.

Maezel demonstrated a talking doll at the 1823 Paris Exhibition, and later took out patent papers (dated 31 January 1824) for the artefact; it is thought that there are no surviving examples. In a survey of London toymakers made by Mayhew in 1850 a maker of talking dolls was found, a tradesman in High Holborn. He is quoted by Mary Hillier (1976) as saying:

> I am the only person who ever made the speaking doll. I make her say 'papa' and 'mama' . . . . The invention of the speaking doll took me many experiments and much study. The thought first struck me one day on hearing a penny trumpet – why not make a doll speak? Science is equal to everything . . . . Many doll makers have dissected my doll to get at my secret. I know of one clever man who tried twelve months to copy it and then put his work in the fire. I laugh and don't care a fig. I have the fame and the secret and I will keep them . . . .

Following Maezel another dollmaker, Alexandre Nicholas Theroude, took out a detailed patent for a talking doll on 30 July 1852: by this time there were real developments – Theroude's device not only managed 'mama' and 'papa' but also 'cuckoo', hardly a talented conversationalist but signalling things to come. Further patents followed in what was becoming an expanding industry. In 1855 Marie Cruchet patented a talking mechanism for dolls, and Jules Nicholas Steiner patented an elaborate talking doll that could also move its mouth and eyes and swing its arms and legs (this figure, like a toddler, needed a helping hand to walk across the floor).

After Thomas Edison had invented and patented the phonograph, a simple talking machine, the way was open for dolls with richer vocabularies. A number of firms produced talking dolls that relied on phonographic principles, most of them dating to the end of the nineteenth century and the beginning of the twentieth. It is obvious that modern dolls are capable of quite complex speech patterns, realized through electronics, but in today's technological climate artificial speech capabilities – where electronic pulses can generate speaker outputs in the form of words – are more commonly incorporated into robot systems (as, for example, with the Omnibot robots from Tomy UK Ltd).

Many walking and speaking dolls have relied upon the stability provided by a wheeled barrow or by walking sticks, and it is common for the activating machinery to be hidden by voluminous skirts. Clockwork mechanisms could be adapted to operate arms and legs, and the head could be made to tilt and the eyes to move from one side to the other. A French doll dating to the first quarter of the nineteenth century moves her head from side to side as her hands are moved to strum a mandoline. American walking dolls of the nineteenth century could either push prams and other small vehicles, or walk unaided across a level surface; and in 1871 artificial crawling babies were immensely fashionable. In that year Robert Clay patented a crawling doll that was later manufactured by the National Toy Company. Such figures could be propelled by clockwork, hidden cords and other elements, methods that could also be adapted for swimming or rowing dolls (and for artificial swimming frogs and dogs). In 1876 the Parisian Charles Bertran designed a clockwork swimming doll which came to be produced as Ondine by the French automata maker Monsieur Martin. The pretty Ondine used the breast stroke (and could also swim on her back), her cork body suitably buoyant and her machinery protected by a metal casing. When the sewing machine was invented in 1830 it was not long before dolls were being designed to use it; one celebrated doll-cum-sewing machine was patented in 1892 by

A. Sandt – the figure operated a handle to control the shuttle and held the needle in its left hand.

It became increasingly desirable for dolls to be lifelike, not only in their appearance but also in their behaviour: eyes that can close and articulated limbs that can assume a variety of humanlike positions have long been familiar, but there have also been many other requirements. In 1867 the French company Léon Casimir Bru patented a baby doll that would cry when a simplified voicebox was squeezed, and also a double-face doll that could show either a laughing or crying visage when a ring on top of the head was turned. The company also produced the Marie-Jeanne doll that could be fed via the mouth into a bladder reservoir that was then emptied by means of a tap. And in 1879 the firm patented a nursing baby doll (named Bébé Teteur), followed in two years' time by an artificial infant that could be fed (Bébé Gourmand). A doll patented in 1892 could talk, sleep and breathe; and 1895 saw the launch of a doll (Bébé Baiser) that could throw kisses. Another kissing doll was marketed by the Steiner company in 1897, here using a counter weight to control its talking valve. While the doll was in an upright position, a pull on a cord would yield 'papa', 'mama' and a kiss; when prostrate, with the cord pulled, a continuous crying sound was produced.

In this field, as in every other, developments in technology came to affect not only the methods of manufacture but also the type of products that were deemed suitable for a new age. The doll tradition of the eighteenth and nineteenth centuries survived securely enough in the twentieth – there is no shortage of walking and talking dolls in the shops of the developed world – but the commercial doll sector has been supplemented by a new and expanding class of artificial beings, the toy robot, an obvious manifestation of technological influence. Today toy robots are commonplace throughout the developed world. They may be simple tin creatures with swinging legs, an encapsulated battery and a few flashing lights; or they may be sophisticated – and expensive – high-technology products with on-board computers and high-level programmability for flexible behavioural response. Once again Japan, part of the worldwide 'coming robotopia', is a good example of the developing robot culture in action.

The Japanese began producing toy robots almost fifty years ago: today specimens that date back to the late 1940s are held in the Tin Toy Museum in Yokohama. Examples of Atomic Robot Man, for instance, are about 5in (12.5 cm) tall and include a wind-up mechanism to enable an uncertain shuffle along an even surface. This fellow ('made in occupied Japan' soon after the Second World War) was manufactured for export, not least to American children who were becoming increasingly sensitive to hi-tech possibilities. Models of the robot Robby, from the feature film *Forbidden Planet*, are also common in the Tin Toy Museum, as is a range of increasingly sophisticated robot systems that are battery-powered and remote-controlled. The owner of the museum, Teruhisa Kitahara, has declared in his book *Wonderland of Toys: Tin Toy Robots*: 'Japanese industrial robots are currently grabbing the world's attention, but between 1955 and 1965, Japanese toy robots had already surpassed others in technology.'

By the 1960s there were already moves to fabricate Japanese toy robots in plastic and other new materials: tin was out. The new products were skilfully designed, efficiently produced and well marketed: they were designed partly for the domestic Japanese market and partly for export. And inevitably they exploited the unique and fertile robot culture in the Japanese tradition. Thus in 1974 the Bandai company, Japan's leading robot toymaker, began producing a toy robot based on Mazinger Z, the giant robot warrior created by artist Go Nagai and already immensely popular through comics and animation. The robot was supposedly made out of *chogokin* (or 'super-alloy'), an advertising ploy – based on the Mazinger Z tales – that brought considerable commercial success. Other robot products followed the same path: for example, the Super Electro-Magnetic Robo Combattler V combined five different robot toys into one; and the 'machine-robo' vehicles – cars, trucks, helicopters – could suddenly convert into robot systems. A wide range of robot-type artefacts were marketed to great effect: by 1984 the Mobile Suit Gundam plastic dolls, precision-scale model robots, had sold to the tune of 100 million, not far short of the total Japanese population at that date. In association with Tonka in the United States, Bandai marketed millions of its Gobot vehicle-robots in America: robotopia is spreading across the globe. Takara, a Bandai rival linked to the US toy manufacturer Hasbro-Bradley, gave a further thrust to these developments by marketing, with immense success, its robot 'transformers' in America and elsewhere.

The robots of the 1980s have been perceived as third-generation systems: toy robots were made out of first tin, then smooth plastics, and finally they exploited advanced 'mechatronic' techniques. In the third generation toy robots – like their large industrial counterparts – may have senses, be capable of speech handling, and be able to perform a wide range of tasks with flexibility and intelligence. Thus in 1983 Tomy created Ki-ku-zo ('I hear', in Japanese), marketed as the Verbot robot in the United States; this toy can recognize up to eight human words as instructions for types of behaviour (stop, start, pick up and put down an object, turn, smile and so on). Today's Tomy robots can sharpen pencils, remove eraser shavings and perform many other humanlike actions. Other Japanese toy robots can nod or shake their heads, or run around exuberently until they are admonished, whereupon they hang their heads in shame.

'Living dolls' and toy robots carry the dream that it is possible for human beings to create humanoid artefacts that are capable of simulating or duplicating human behaviour. Throughout the centuries the humanoid automata were nothing more than a metaphor for human behaviour in general and human thought in particular, but today the metaphor is being translated into reality – the artefacts are acquiring intelligence and life characteristics. The simple automata, the toy robots and all the attendant myths can now be seen as the true heralds and ancestors of artificial life. In particular, for the purpose of this book, they prefigure the anthropomorphic systems, the proliferating robot species, that are so remarkable a feature of modern computer-based technologies. Before taking up the theme of robots proper it will be useful to look at another evolutionary thread that has contributed to the modern robot tapestry. Rudimentary thinking machines show how purely mechanical thought was first built into artefacts.

Mechanical Thought

Few of the historical automata could claim to have much to do with actual thought. the cerebral activities that characterize *Homo sapiens* and many other biological species. The artificial dolls, the artificial musicians, the artificial warriors – all gave the appearance of being able to think, of being able to decide, being conscious, but only the most superstitious people imagined that such artefacts had a real mental life. Most historical automata were not only mindless; they lacked the seeds out of which artificial minds might evolve in the centuries ahead. But at the same time there was a small group of machines – rare in the ancient world, more common after the Middle Ages – that focused on the processes of reasoning, on some of the procedural tasks that nineteenth-century mathematicians began to call the 'laws of thought'. In short, there was one particular family of historical automata – the mechanical calculators – out of which the artificial brains of the modern age would evolve.

There were few mechanical calculators in antiquity, though many methods of calculation were developed. Babylonian clay tablets and Egyptian papyri carry details of how numbers can be manipulated to achieve particular results. For example, the Senkereb Tablet, found near the site of ancient Babylon and about four thousand years old, carries cuneiform numbers set out in rows and columns to aid the task of computation. Similarly, the Babylonian Table Texts were used to help perform practical calculations in such areas as calendar compilation, agriculture, stock control and military logistics. Mathematical innovations are indicated in the Egyptian Ahmes Mathematical Papyrus, four thousand years old and today held in the British Museum; in the Hindu Sulva sutras; and in many historical Arab texts – we owe the words 'algebra' and 'algorithm', along with many mathematical advances, to the Arabs.

The first mechanical calculators were bones, sticks and stones on which were scratched linear patterns to represent numerical quantities. Some observers interpret Stonehenge and the lesser henges as astronomical computers, positioned with care to signal particular astronomical events. Less controversially, the abacus, invented in ancient China and still in widespread use today, is an obvious example of a mechanical calculator. The Romans and Greeks acquired types of abacus, but then also made attempts to develop calculating machines based on new principles. Thus Greco-Roman hodometers, akin to primitive mechanical milometers, were used for simple numerical operations. And it is also worth mentioning the Antikythera machine, an interesting example of what might be regarded as an ancient computer . . . .

Today the National Archaeological Museum of Greece in Athens contains corroded fragments of a strange metallic device found by sponge divers off the island of Antikythera in 1900. The object, dated to around 65 BC, contains complex dials, gears and other elements. Until 1959 it was assumed by the Museum that the device was an astrolabe – an early navigational instrument – but then the scientist Derek J. de Solla Price came to identify the machine as an ancestor of later mechanical calculators. He declared: 'It appears that this was, indeed, a computing machine that could work out and exhibit the motions of the sun and moon and possibly also the planets' (*Natural History*, March 1962). The new interpretation of the significance of the Antikythera device had dramatic implications, signalling above all the remarkably advanced state of technology in ancient Greece. In Washington in 1959 Dr Price commented: 'Finding a thing like this is like finding a jet plane in the tomb of King Tutenkhamen.' In fact there is some evidence of such devices in history: for example Al-Biruni, a Persian traveller in India, described a similar artefact around AD 1000. Mechanical clocks too, with their long history, have been depicted as systems able to compute by counting events.

Another mechanical calculator, 'a kind of primitive logic machine' (Gardner, 1958), may have been invented by the Spanish Franciscan thinker Ramon Lull (1232–1315) to facilitate the use of his logic system, but he had an unusual vested interest. His masterpiece, the *Ars Magna* (the 'Great Art'), was supposedly revealed to him by God, and one of Lull's main purposes was to use methods in formal logic to reveal the supposed 'truths' of Christianity. In this we may assume him to have been unsuccessful, but none the less he had considerable influence. The automatic 'book generator' in Swift's *Gulliver's Travels* (Part III, Chapter 5) may well have been inspired by Lull's methods; and Giordano Bruno, the celebrated Renaissance martyr who had little enough reason to love Christianity, was prepared to regard Lull as 'omniscient and almost divine'. But perhaps the most important Lullian influence was on the German philosopher and mathematician Leibniz, who acknowledged in his *Dissertio de arte*, written in 1666 when he was nineteen, that Lull had pointed the way to a universal algebra, a stepping stone to the computers of the modern age. Leibniz himself developed an important mechanical calculator, a graphic illustration of the tenor of the times.

By the seventeenth century many of the mechanical principles that would aid the construction of calculators had been developed. The engineering of mechanical calculators depended upon intricately organized parts that could move with precision. It was conventional for such parts to be manufactured by clockmakers, the contemporary state-of-the-art technicians. The first clock escapements had been designed and fabricated in the thirteenth century; and coiled springs, used to replace driving weights, were incorporated in portable clocks (the first watches, about the size and shape of an orange) at the beginning of the sixteenth. Further mechanical elements (the 'stackfreed', Jacob Zech's 'fusee' and so on) were developed to maintain the force of the spring constant as it unwound. Such inventions provided the body of practical technology that was necessary for the construction of the first realistic calculators.

In the fifteenth century, financial transactions were computed by bankers and other dealers using the convenient squares of a tablecloth (resembling a chequerboard, hence the term 'cheque', just as the abacus beads called 'calculi' give us the word 'calculate'). At the same time various thinkers were wondering how to devise machines that may be used to aid the task of computation. Leonardo da Vinci, for example, in addition to other preoccupations, was considering how to build a mechanical calculator; there is no doubt that had he lived a century later the available technology would have been much more conducive to such ambitions. In fact one of the first efforts to build a practical calculator was made by the Scottish mathematician John Napier (1550–1617), of logarithmic fame (here simple trains of gear wheels were supposed to replace the more primitive Napierian 'bones'), but the first truly significant device was developed by William Schickard (1592–1635), a thinker with wide interests in logic and mathematics. Schickard's was lost in a fire, and the existence of his 'calculating clock' is only known about by the reference to it in a letter to the astronomer Johannes Kepler in 1624.

Another mechanical calculator was constructed by the French philosopher and mathematician Blaise Pascal in 1642 (one of the machine's copies, today in the Paris Conservatoire des Arts et Métiers, carries Pascal's signature). He offered his calculating device with the words: 'I submit to the public a small machine of my own invention by means of which alone you may, without effort, perform all the operations of arithmetic, and may be relieved of the work which has often times fatigued your spirit.' The machine, known as the Pascaline, could only be used for addition and subtraction (multiplication being carried out by repeated addition), though perhaps his was remarkable enough for the times. To add numbers, the first was dialled in, and then the next and so on; the result appeared at windows in the machine. A decimal transfer provision was included by arranging for a gear wheel to move on one unit; when a wheel was moved from 9 to 0 a lever, lifted by pins and fastened to a pawl, fell downwards, causing the next wheel to move one step forward. Numbers were set on drums for display at the windows.

Pascal was aware of the limitations of the machine, and in fact thought about how superior devices, based on different principles, could be built: in 1685 he described plans for an upgraded Pascaline based on the use of a variable-toothed gear operating with a system of pulleys. At the same time he knew that the 1642 machine was a significant step forward: his sister observed that his 'mind had somehow been taken over by the machine'. A device similar to the Pascaline was constructed by Sir Samuel Morland in 1666 (Samuel Pepys commented: 'Very pretty but not very useful'). Leibniz addressed himself to the weaknesses of the Pascaline and in due course managed to construct a machine able to add, multiply, divide and calculate square roots. Here stepped rolls, rather than gear wheels, are employed, with use also made of a cog-wheel, a square shaft, setting devices and a scale. The operator only had to turn a crank and the resulting number could be read in the windows of the product register. Other mechanical calculators were invented by Lepine (1725), Hillerin de Boistissandeau (1730), Jacob Pereire (1751), Poleni (1709) and W. T. Odhner (1891) among others. In 1770 Jewna Jacobson, a clockmaker at Minsk in Russia, invented a machine that could handle numbers up to five digits; and throughout the eighteenth and nineteenth centuries a host of mechanical and theoretical innovations contributed to the development of truly autonomous calculating machines. In 1801 Joseph Marie Jacquard (1752–1834) invented the automatic drawloom, in which punched cards were used to control the lifting of thread and the consequent generation of the fabric pattern; like Pascal, Jacquard was, on his own admission, 'obsessed' with his research. Today the original Jacquard loom is seen as, in effect, a mechanically programmed computer, the first device to store a program and control a machine.

By the early nineteenth century, engineers were equipped to supply a host of precision components for mechanical systems: cogs, levers, dials, drums, toothed wheels, ratchets, spined cylinders and so on. And there was a parallel evolution of the number theory and symbolic logics that would be essential for the electronic computing of the twentieth century. The scene was set for new advances in mechanical computation: many of these were accomplished by the English mathematician Charles Babbage (1791–1871), whose achievements, in practical terms, have often been depicted as disappointing but which none the less laid the basis for many later computer developments.

Babbage, sometimes dubbed the 'father of modern computing', helped to inaugurate the London-based Analytical Society to discuss theoretical questions largely of a mathematical nature; by 1821, aged thirty, he had written five books, nearly twenty papers and many other items. With J. F. W. Herschel he was also instrumental in setting up the Royal Astronomical Society, and in consequence was obliged to compile extensive reference tables, much to his chagrin. When Babbage exploded with the words, 'I wish to God these calculations had been executed by steam!', Herschel commented that it should be possible. It was around this time that Babbage wrote to Sir Humphrey Davy, president of the Royal Society in London, proposing the development of a machine to replace 'one of the lowest occupations of the human intellect'. It seemed that Babbage was unimpressed by the Pascaline, the Leibniz machine and all the other existing devices intended to ease the task of computation.

Babbage first developed the Difference Engine, a machine intended to calculate and check mathematical tables. In one version the device was able to work to six decimal places, but ambitions for a more complicated machine were thwarted by production problems. A portion of the new Engine was constructed by Joseph Clement in 1832, and now resides in the London Science Museum; while in 1859 a version of the machine was built to calculate life tables for insurance companies. But by the early 1830s Babbage had acknowledged the limitations of the Difference Engine and began to develop of a new mechanical calculator based on other principles. Then, as now, there were problems in obtaining funding in Britain to support research and development. In 1835 Babbage wrote to an American: '*You* will be able to appreciate the influence of such an engine on the future progress of science – I live in a country which is incapable of estimating it' (quoted by Hyman, 1982).

The main task confronting Babbage now was to construct the Analytical Engine, an ambitious system that would suggest the enduring architectures of twentieth-century electronic computers (the best descriptions of the Analytical Engine are largely due to the efforts of Babbage's colleague Ada Lovelace). Hyman has suggested that in 1834 Babbage saw 'a vision of a computer, and remained enthralled for the rest of his days': he worked until his death on the realization of a practical computer system, falling short of his ambitions but laying the basis for the computer architectures of the future.

Before the middle of the nineteenth century – a hundred years before the invention of the transistor and the emergence of modern electronic computing – Babbage understood that computers would have to include five key components or facilities: input, to allow numbers to be fed in; store, to hold numbers and program instructions; arithmetic unit, to perform the actual calculations; control unit, to provide overall supervision of task performance; and output, to communicate results to users. Such components can be found in all modern computers. When Charles Babbage died in 1871 his youngest son Henry struggled to manage the daunting legacy of the mechanical computational engines, but there were few further developments in the theory and practice of these formidable machines. It had to be acknowledged that the necessary enabling technologies were not available: cogs and levers cannot be induced to support high levels of computation with any efficiency, however sound the architectural insights and the mathematical theory. In 1879 a committee was formed to stimulate developments in Babbage's work, but little was accomplished. Much of the computing theory was in place but a new technology – to aid practical realization – was needed. In due course, after nearly half a century, this would be provided by electronics.

Enter Electronics

In the 1920s and 1930s a number of mechanical and electromechanical systems were built to perform computational activities. The ASCC (Automatic Sequence Controlled Calculator) was one such electromechanical monstrosity; finally completed in the early 1940s, this computer was 50 ft (15m) long and 80 ft (25m) high, unreliable and difficult to maintain. It relied extensively on electromechanical relays, a preferred alternative to the then new-fangled and extremely unreliable thermionic glass valves, to form the basis of the system logic. In fact it was the glass valve, a component in many early computers, that was to point the way to the electronics future.

The thermionic glass valve, much used in early radios and television receivers, is essentially an evacuated glass bulb carrying metal electrodes; voltages across the electrodes control the currents flowing in the vacuum. A heated electrode emits electrons, allowing a current to flow. Thomas Edison was the first researcher to observe thermionic emission – at the Menlo Park laboratory in 1883 – and so, for a while, the phenomenon was dubbed the 'Edison Effect'. When Edison could not think of a commercial application he lost interest ('Well, I'm not a scientist. I measure everything I do by the size of the silver dollar. If it don't come up to that standard then I know it's no good'). Further research, by other people in other places, revealed that the glass valve could be used in many applications; for example, to amplify radio signals and to serve as a two-state switching device. It was this latter use that was of particular relevance to the development of modern electronic computing: fast two-state switching devices can be used to realize the binary logic* on which most practical computing systems are based.

In a parallel development it was being appreciated that various semiconductors (materials such as silicon and germanium, poor electrical conductors) had interesting electronic features. It was found in due course that semiconductor materials (typically silicon and germanium), sometimes 'doped' with other substances, could perform useful electronic tasks that could be applied in many different ways. In particular, for the purposes of computer development it was found that semiconductor-based transistors could perform many of the tasks of the traditional glass valve, but much more cheaply and much more reliably. Thus the first generation of electronic computers based on unreliable glass valves were quickly supplanted by a new generation based on transistors. Later generations of computers were based on further refinements to semiconductor technology. Ways were found to build thousands of circuits into minuscule silicon chips, a development that led to endless application scenarios. For example, compact computer brains could be incorporated in other manufactured systems – to control their behaviour and vastly to increase their flexibility of response. One key application in the modern world is to use silicon-based intelligence, in either free-standing or on-board computers, to extend the versatility and competence of the robot.

Many types of automata have been described in this chapter. Examples have been taken from different times and different cultures to indicate the ubiquity of automata theory and automata practice throughout history and throughout the world. Practical automata, like the myths and tales surveyed in Chapter 1, have helped to keep alive the idea that human beings can design humanoid artefacts and translate the design into practical realizations. The auto-

* Since we are most familiar with the 0 to 9 ('decimal') number system, it is not always appreciated that all computations can be carried out using only 0 and 1 (i.e. using 'binary' methods).

mata have been built to every imaginable level of complexity, from the simplest systems of the ancient world to the most complex hi-tech creations of the modern age; from the primitive movable-limbed dolls of prehistoric societies to the computer-based toy robots conceived in modern design offices and born on automated assembly lines.

It has been useful to highlight two broad classes of automata. On the one hand are the simple anthropomorphic specimens – the artificial musicians, warriors, scribes, babies and so on – intended to resemble human beings in everyday roles and activities; and on the other are the calculators, the number-manipulating automata designed to simulate or duplicate a particular human organ, the brain. The two broad classes of automata have remained distinct through much of their historical evolution: there is no evident relationship between an ancient doll with articulated limbs and a Greek hodometer, between an artificial baby able to cry and a mechanical calculator able to multiply and take square roots. Yet the two types of automata have experienced a dramatic convergence in the modern age: the humanoid artefacts have acquired working brains. This singular fact has helped to define the modern computer-based robot. For the first time in history human beings are able to frame artefacts that in some important sense are replicas of themselves: the age-old dream of the mythmakers, occultists and *karakuri* masters is at last being achieved.

To some extent the terms have been loosely defined, the demarcation lines loosely drawn. This chapter began by indicating some traditional definitions of automata; a key feature was the idea that automata are self-moving. Yet most of the artefacts described are not capable of any independent motion: the humanoid specimens are reliant upon rain or wind or human energy; the mechanical calculators depend upon numerical quantities injected by human beings, upon dials and levers operated by human. If we are being semantically rigorous it has to be admitted that historical automata are rarely, if ever, truly autonomous. Instead we are encouraged to take an imaginative leap; when the artificial musician begins to play, we are encouraged to suspend our critical faculty, to take 'as if' to be as good as the real thing, to equate mimicry with duplication. It is part of the fascination of the modern robot that the imaginative leap is becoming increasing unnecessary. The modern robot is evolving its own intelligence, its own mental faculties set in a humanoid frame. Now let's begin to consider the character of robotics in the modern world.

3 Modern Robotics

There are many ways in which modern robots are complicated. For a start, the cleverest among them are trying to behave like human beings – which is no mean trick. Such robots – suitably anthropomorphic in appearance – come equipped with many of the anatomical bits and pieces that characterize a typical member of the species *Homo sapiens*. They can, for instance, have a torso, arms, hands, legs, eyes, ears, nerves and a brain. They do not have many of the mammalian organs – such as a spleen, heart or kidneys – that are useful to human beings, nor do they need them: the modern robot is evolving its own characteristic ways of handling energy and information, and because robot evolution is guided by human and machine intelligence, rather than by biological trial and error, the process is speedy and efficient.

The evolution of robots has already generated many different artificial species (see next section), which variously exploit the available robotic options; but many of the functional systems derive from the same evolutionary roots, the original innovations that began to lay the basis of robot theory and practice. There is some dispute about who designed the first industrial robot. The United States makes a number of claims based on the achievements of Unimation, today a leading robot manufacturer. But it is also interesting to remember the efforts of a British engineer, Mark Vale, who in 1946 designed a twin-arm robot to pick washers from a chute and to place them alternately in two containers (see the brief account of this machine in Astrop, 1983). The system operated electromechanically, under a sequence controlled by a slowly revolving drum that carried trip dogs to actuate limit switches. Articulated arms, familiar in modern robots, were used, and functional hands ('grippers') were controlled by cables. This robot was what came to be known as a 'pick and place' device, a robot of the first generation. It was able to pick switch knobs from an inclined chute, apply them to a polishing bob, and then place them in a tray; at particular preset intervals the robot would apply fresh compound to the bob.

The original system could only handle a few ounces, but later models were designed to handle payloads up to 8 lb (3.6 kg). By the mid 1950s Mark Vale was in business with the company Gilmour & Vale in Wellingborough, Northants, and systems were being commercially produced – though with little success. A Vale robot set up in 1956 as a process-serving unit evoked little interest in the outside world (Mark Vale himself remarked that 'though sales representatives tried hard to obtain enquiries, nobody showed interest. We therefore continued to use it ourselves'). The project was dropped, then resurrected in the late 1960s when a larger robot was designed. Thus the Vale initiatives yielded a number of practical working robot systems (none of them commercially successful):

1946 Twin-arm 'pick and place' model
1948 Single-arm version with cable-controlled gripper
1956 Automatic system able to handle 8 lb (3.6 kg) payloads
1975 Two-speed drive unit covering 30x34x24 in (76x86x61 cm) working area

By the 1970s the Vale systems were being rapidly and prolifically outflanked by more sophisticated designs from the United States, Japan, Sweden and elsewhere. The scene was set for a massive explosion in the world's robot population.

Another robot pioneer, the American George C. Devol, had forty patents to his name, one of which, taken out in 1954, was for 'Program Controlled Article Transfer', an important robotic innovation. Devol showed the design to Thomas J. Watson Sr, the president of IBM, who at the time was unable to interest his colleagues who were busy developing the mainframe computer business. Devol, quoted by Asimov and Frenkel (1985), commented: 'They had their hands full many times over and "Why do we want to go off in another direction?" was the attitude.' But when Devol met Joseph F. Engelberger at a cocktail party he obtained an enthusiastic response: Engelberger later founded Unimation Inc., and was excited about being involved in robot production. Interviewed in 1983, he said: 'Over and over, the advice was, "Don't call it a robot. Call it a programmable manipulator. Call it a production terminal or a universal transfer device." The word is *robot* and it *should* be *robot*. I was building a robot, damn it, and I wasn't going to have any fun, in Asimov terms, unless it *was robot*. So I stuck to my guns' (emphases added in transcript).

By 1964 Unimation had sold thirty industrial robots. Today more than ten thousand of its systems are installed around the world. Unimation remains, within a select group of American, European and Japanese companies, a lead player in the massive international robot business.

What is a Robot Today?

History has seen many devices and myths that today may be termed robots – from the mobile tripods in the *Iliad* to the mechanical moving elephants of Indian legend. In modern technology, 'robot' is an almost equally loose term. For our purposes it vaguely resembles a human being in having limbs, torso, sense organs and the like. Or, as in one early book on robots (Strehl, 1955) and much current usage, the term may signify nothing more than a box of electronic components, as with the 'robot pilots' used in aircraft.

According to the US Robotic Industries Association (RIA), formerly the Robot Institute of America, an industrial robot is a reprogrammable, multifunctional manipulator designed to move materials, parts, tools or specialized devices through variable programmed motions to perform a variety of tasks. The important facility of reprogrammability implies the provision of computer control (see Chapter 4), and distinguishes 'true' robots from simpler machine tools and other modern manufacturing devices. A definition favoured by the Japanese Industrial Robot Association (JIRA) encompasses a wide range of systems, including relatively simple robots that cannot easily be reprogrammed to perform fresh tasks (for this reason, statistics for Japanese robot populations are sometimes inflated). Where a robot is easy to reprogram it may be dubbed 'soft' or 'flexible'; a simple robot unable to learn new behaviour is often regarded as an example of 'hard' automata.

In the approach adopted by the British Robot Association (BRA), a true industrial robot must be fully programmable, in contrast to the simpler 'pick and place' devices of which in the past there have been many more. For example, by 1980 Japan had seventy thousand industrial robots, of which only about 5 per cent were fully programmable (or 'reprogrammable'). Engineers at General

Motors in the USA have defined an industrial robot as a 'programmable device capable of performing a complex series of motor actions over a wide variety of processes', and as a 'reprogrammable multi-axis mechanical manipulator'. Such definitions would be expected to include hydraulic robots normally used for such operations as materials handling and spot welding, pneumatic robots used for small assembly tasks, and electrical robots used in a wide variety of industrial applications.

It is also possible to classify robots according to features and facilities other than those relating to (re)programmability. For example, a typical industrial robot may be regarded as comprising three elements – mechanical structure, control system and power unit – which may be used to help define the class of system.

The mechanical structure typically consists of a central pedestal, which may be rigid or articulated, static or mobile, free-standing or mounted on wall or ceiling. Any movement of the central pedestal may be supplemented by movement of an extended articulated arm that terminates in a tool such as a welding torch or drill, or in a gripper able to manipulate different tools as required. Where the arm carries a wrist and gripper, three further axes of motion can be added, the total number of axes being identified as the number of degrees of freedom – an index of the robot's versatility.

The control system, too, helps to define the robot type. There are many types of possible controllers: pneumatic logic systems, electronic matrix boards, electronic sequencers, minicomputers and microprocessors. After the rapid developments in semiconductor technology through the 1980s, most high-level robotic systems are today controlled by solid-state circuits, silicon chips able to carry hundreds of thousands of electronic components. In one early interpretation (Tanner, 1977), robots are either non-servo or servo-controlled. Non-servo robots use fixed or variable mechanical stops and limit switches for positioning purposes and to inform the controller. Servo-controlled robots employ sophisticated feedback devices and monitor the appropriate variables until they reach a point at which the controller is instructed to stop actuating the robot arm and tool. It has also been useful to classify servo-controlled robots as either point-to-point (PTP), where there is no control of the path between the points to be reached, or continuous path (CP), where path control achieves a continuous, well-defined movement.

The power unit, required to activate the entire robot system, may be hydraulic, pneumatic or electrical – or a combination of all three. Most modern industrial robots have electrical and electronic features.

It is sometimes said that robot classifications based on type of mechanical structure, control system or power unit are, in an important sense, 'static', in contrast with 'dynamic' classifications based on robot behaviour and performance capability (Engelberger, 1976). With the dynamic approach the focus is on such aspects as load capacity, acceleration, deceleration, speed, working conditions, manipulative capability, ease of programming, reliability, sensory facilities and the capacity for self-diagnosis. It is obvious that the more advanced robots become, the more scope there is for distinguishing between different types.

The 'generations' of robots are defined according to both static and dynamic criteria. The simple pick-and-place devices of the first generation have already been encountered. Subsequent generations are recognized according to the capabilities of the systems, realized through such elements as vision facilities, the capacity to coordinate received information from different senses, and the ability of the robot brain – as either a linked computer provision or an on-board component – to make judgements, to take decisions and to derive conclusions using in-built expertise in one or more disciplines. Most industrial robots belong to the first two generations; later generations encompass the relatively few highly advanced systems, research models, and the theoretical concepts floating through the minds of design engineers.

There are more industrial manipulators, of one generation or another, than any other type of robot in the modern world. However, many other robot species, apart from simple boxes of electronics to offer control facilities, deserve attention. Apart from industrial robots there are:

- *Medical simulators* Figures designed to mimic medical conditions, healthy or otherwise
- *Exoskeletons* Devices to which human beings can link themselves to amplify natural physical abilities
- *Autonomous guided vehicles* Mobile systems, normally used in factories, that are capable of self-navigation
- *Remotely controlled vehicles* Suitable for bomb disposal or work in other types of hazardous environment
- *Programmable mannequins* Figures that may be programmed to carry out obvious human roles (receptionists, waiters, etc.)

- *Toy robots* Humanoid artefacts of growing sophistication (current models have conversational, manipulative and self-navigational abilities)
- *Domestic robots* Capable of carrying out simple household chores, but as yet mainly of entertainment and educational value
- *Promotional robots* Used at product exhibitions and in advertising films
- *Space probes* Sophisticated systems able to self-navigate and feed material samples into an internal analysis laboratory

One robotics specialist, Professor M. W. Thring, defines a robot as 'a machine with a computer memory which can be programmed by a human to carry out any one of a variety of different series of complex manipulative tasks . . .' (Thring, 1983). He also identifies two other classes of systems: mechano-chiropods (exoskeletons and prosthetic devices for the handicapped), and telechirs (strength amplifiers, devices for use in dangerous places, and micromanipulators and surgical devices). Thring uses the word 'sceptrology' (based on a Greek word that denotes the king's staff of power and also a crutch) to refer to the class of artificial limbs designed for handicapped people. 'Telechirics' (literally 'hands at a distance') denotes systems controlled at a distance by human operators using communication links. Examples are remote mechanical hands working in a radioactive environment, and minuscule devices carrying out surgery inside a human body.

It is clear that all definitions and classifications are arbitrary, though they are useful where a consensus has emerged. Thring identifies four classes of robots (senseless, immobile; partially sensed robots; mobile robots; and fully sensed mobile robots), indicating that the more advanced systems are capable of using sensory inputs to adapt to their environments. There are three broad areas into which robots fall:

1 Robots may be regarded as active systems in the real world. They are intended to be engaged in well-defined tasks in industrial, commercial, domestic or other environments. They often rely upon computer intelligence, but are necessarily interested in more than mere cogitation: they are interested in physical behaviour in the real world.
2 Not all robots are anthropomorphic. They do not all have torsos, limbs and senses. There are plenty of robot systems that are well applied to specific tasks, without any physical

resemblance to human beings or any other biological systems. Perhaps humanlike activities are performed by non-humanoid systems, or perhaps totally unprecedented modes of behaviour are accomplished. (However, much of what follows in this book focuses on anthropomorphic robots.)

3 Robots become more competent – in terms of both specific tasks and flexibility – as they acquire and evolve sensory faculties and improved intelligence. In this way robot evolution closely parallels evolution in the biological world: 'superior' species are recognized by the enhanced acuity of their senses and by their highly developed cerebral equipment. The 'ultimate robot' will be a creature with superb senses, excellent mobility and a supreme mind (doubtless a fellow far beyond mere mortals).

This book will discuss robots that have humanoid characteristics, in terms of both their physical architectures and their mental capabilities. It should not be surprising to find that a substantial class of modern robots are evolving in the direction of *Homo sapiens*: human beings are, after all, an extremely successful species (perhaps too successful). Artificial systems with an interest in having an impact on the world would be inclined to develop humanoid characteristics. Let's begin by taking a look at artificial anatomies, at the quasi-biological bits and pieces that in due course will enable robots to emerge as persons in their own right.

Artificial Anatomies

The various mechanical, fluidic, electrical and electronic components that together comprise the modern sophisticated robot may reasonably be regarded as the biological anatomy of the device. The central turret or pedestral of the typical industrial robot may be regarded as the torso (with an abdomen that may well conceal many 'vital organs'). As with many biological species, the torso provides a robust structure from which the attached arms can operate (in principle, a robot can be designed with one arm, two arms or more). Biological eyes are typically situated in a creature's head, but there is much more variety in the design of robot systems. It is not uncommon for robot eyes to be situated in the grippers – an eyeball in the palm of the hand (one early example of this is described by C. Loughlin in his paper 'Eye in hand robot', delivered at the Second International Conference on Robot Vision and Sensory Controls, November 1982). There is similar flexibility in the positioning of robot brains. It may keep its brain in another room, yards away from the rest of its body, or in its abdomen, or it may have several small brains, variously allocated to its eyes, its ears, and the joints of its limbs. So there are areas of design flexibility with robot sytems, in some cases far exceeding what can be found in the natural biological world, But it is also obviously true that biological systems have the edge in many particulars.

Many insights into robot design have been gained by examining the anatomical structures of biological systems; this was a familiar ploy of the ancient automata makers, the *karakuri* masters and others. One example of this approach is how some robot fingers are designed (see next section); another is the interesting design of an artificial spine (which in fact is applied to serve as a unique robot arm).

In 1983 a novel design for a painting robot was announced by the Swedish firm Spine Robotics of Molndal. Robots usually have arms similar to their biological equivalents – a number of rigid members articulated at joints. The novelty of the new design from Sweden was that the system was structurally similar to the human spine, The two arms (spines) of the robot each carry a hundred disks connected together by four steel cords; hydraulic actuation is used to pull the cords in a precise way to move the spinal arms.

The new system was seen to have a number of advantages over conventional robot designs. The arm has a long reach (up to 12 ft (4m) high and 8 ft (2.4 m) in any horizontal direction through 360 degrees). The 'working envelope' of the robot is almost a hemisphere, and its flexibility is such that a tool can be held stationary at the end of the arm while the rest of the spinal arm is able to move. The arm can therefore bend round corners and access difficult places that it would be quite impossible for the human arm to reach. The basic idea was conceived by Ove Larrson in the mid 1970s, and mathematical colleagues at the Swedish Chalmers University worked out the algorithms that were necessary for automatic control. Research projects linked to the artificial spine have yielded well over a dozen Masters degrees, and the robot has been tested in many practical implementations (for example, in manufacturing applications at the Volvo motor vehicle organization).

Another 'biological' design for robots concerns the building of artificial muscles. Scientists at Hull University in Britain have discovered a way of producing robot muscles that could operate by the expansion and contraction of polymers (*New Scientist*, 5 November 1988). Since such artificial muscles can be made to stiffen or relax simply by adding a chemical, they are seen to be much more compliant than those driven by current techniques. Again, the idea derived directly from a study of normal biological methods: many living creatures convert chemical energy into mechanical energy to achieve movement, a strategy rarely used in artificial systems. The researchers focused on a co-polymer of polyvinyl alcohol and polyacrylic, first studied in the 1950s in connection with work on muscle behaviour.

The scientists at Hull have produced artificial muscles that produce gripping forces similar to those achieved in the human hand. Two complementary muscles, a flexor and an extensor, have been attached to opposite sides of a gripper. Each muscle comprises a watertight chamber containing strips of polymer muscle that are activated when appropriate chemicals are added to the chambers. And the artificial muscles are under intelligent control: a computer is used to control the quantities of chemicals that are added. There are similar research projects elsewhere. For example, Daniel De Rossi at the University of Pisa in Italy is using electrolysis to control the acidity of the muscle chambers, while the Japanese are using electricity to stimulate polymer muscles.

Such examples show how scientists are taking advantage of biological knowledge to further research into robotics. It is also practical to adapt new technologies to the development of robot systems: one clear example is the way in which optic fibres can be used as artificial nerves to carry pulsed information around artificial limbs and other parts of a robot's anatomy. Thus optic fibres, which are of immense importance in communications technology, can be used in the context of the way biological systems distribute information; and so the manipulation of information that is essential to robot intelligence can be enhanced.

Learning from biology, applying new technologies in an opportunistic way, developing specific programmes of robotics research – all such ploys are necessary to aid the progressive evolution of robot systems. As sophisticated robots develop in the direction of *Homo sapiens* their debt to biology and their own claims to a biological status become equally evident.

Give Us a Hand

Many researchers have commented on how difficult it is to design and build an artificial hand. One writer (Pawson, 1985) even suggests that it might be easier to build an imitation of the human brain than an imitation of the human hand. In fact many efforts have been made to design and build articulated fingers that can be used in robot applications, and many other types of robot hands ('grippers') have been produced.

A typical gripper, or 'end effector', may have two opposed fingers designed to grasp workpieces or tools. Other grippers may have three or more fingers, but as the number of fingers increases so do the associated control problems – for example, in ensuring that each finger applies the same pressure to the grasped object. Some hand designs have attempted to duplicate the opposed thumb of human beings and other primates. However, it will seen that humanoid grippers are only one possible approach to the design of robot and effectors. Again it is worth remembering that the construction of artificial hands, and of artificial limbs in general, has a long and interesting history.

As early as 1509 an iron hand was produced for the German knight Goetz von Berlichingen, immortalized in Goethe's play of the same name. This metallic construction, carrying operational gears for fingers and thumb, was of limited use, and Goetz himself commented: 'My right hand, though not useless in combat, is unresponsive to the grasp of affection. It is one with its mailed gauntlet – you see, it is iron!' But despite the limitations of this particular hand it should not be thought exceptional: many prostheses were being designed and fabricated in the sixteenth century, including the devices manufactured by the famous military surgeon Ambroise Paré (1510–90). It was inevitable that such artefacts should be clumsy and difficult to use. None the less they embodied impressive state-of-the-art technology, and helped to advance the theory and manufacture of ratchets, levers, springs and gears. Then, as now, military needs were a powerful engine for technological development. Artificial hands were devised with fingers extended by springs and flexed by clever systems of ratchets and levers. There are clues here for the design of robotic hands in the modern world.

A hand produced by the designer D. W. Collins in 1961 has a metal structure but also includes a soft covering for the fingertips. The aim is to make the device feel like a human hand when shaken by a person. Collins's hand has been produced in various sizes, and is versatile enough to exert appropriate degrees of pressure when grasping objects. A flexible three-fingered hand produced by the Automatic Control Division of the Electromechanical Laboratory in Japan includes elaborate control provisions operated by electrical power. Here the fingers are able to extend, bend and pitch under intelligent supervision. Another hand and arm device from Waseda University, an important robotics centre in Japan, is powered by a hydroelectric system that includes a minuscule oil pressure pump; while the XI hand, made at Chuo University, also in Japan, has five fingers with a delicate sense of touch based on 384 contact points arrayed to allow the hand to recognize the shape, texture and size of objects in its grasp. The Cranfield Institute of Technology in Britain has also designed a five-fingered hand and an operating prototype has already been produced.

It is obvious that for many tasks conventional hands, natural or artificial, have serious limitations. For example, they are not ideally suited to lifting large sheets of glass or heavy bales of straw; and robotic hands, amenable to intelligent design, might be expected to develop various non-anthropomorphic features. In fact there are already many designs for artificial hands that do not resemble anything to be found in the natural world. Sheets of metal, glass or plastic can best be handled by magnetic or suction grippers: a secure 'grip' is easy to achieve and can be made to release simply by the automatic or manual flicking of a switch. Other grippers may comprise an array of spikes to impale certain classes of objects, such as large bales of waste paper, that need to be transported. Human beings have evolved hands, which we all recognize as immensely versatile devices; robots are evolving fingered hands and many other manipulative facilities.

Robot hands can be powered in various ways (see next section), offering a range of provisions for different purposes. Hydraulics can provide high levels of clamping power, but pneumatics offer a more accurate control of the applied forces. Another possibility is 'proportional grippers', under electrical control and able to close to a predetermined position according to the specific needs of the application. Gripper pads, provided to avoid damaging delicate items, may be made of polyurethane bounded to a metallic base. Some fragile objects may be handled by an expandable bladder or balloon, and there is also scope for using end effectors coated with adhesive – though 'release' may be difficult to accomplish in any precisely controlled way.

Another option is to pad artificial fingers with a loose lining filled with granular powder. When the device is activated, the padding is pressured to mould round the object to be grasped, after which the granules can be locked in position by applying a vacuum or an electromagnetic force. Such a device has the advantage that it can accommodate its shape to the object to be grasped, even if the object has never previously been encountered. The granules roll freely, and to some extent are out of control: the robot cannot 'know' with any precision what each granule is doing, and so it cannot infer the shape of the grasped object (it may 'look at' the object, but that is another story – see Chapter 4). However, the granular principle can be used in ways that avoid the actual use of grains; there are obvious advantages in allowing a gripper to flow round a grasped object and to know at the same time exactly what sort of 'flowing' is taking place. To some extent this has been accomplished in the design of the Omnigripper, developed at Imperial College in London.

In this design, each of two parallel fingers carries an array of spaced pins that can move up and down independently of each other. When the device is activated to grasp an irregular object, some of the pins are pushed up to accommodate swellings or protuberances; in such a fashion the pins are positioned to shape around the individual object. The Omnigripper fingers can close to achieve an external grip; or – from a closed position – they can be powered to open in an aperture to accomplish an internal grip. Information derived from the position of each pin can be interpreted to provide details about the shape of the grasped object, so allowing the object to be recognized.

The use of balloons, bladders, granular pads and the Omnigripper together represent the technology of 'soft grippers'. Such a technology may be seen as being essentially 'object-friendly' – designed to be sympathetic to object peculiarities and vulnerabilities. Another type of 'soft gripper' is built out of two tentacles made from chain links, inside of which are pulley wheels carrying thin cords. When tension is applied to the cords by an external electric motor the tentacles are activated to unwind around the object to be grasped. As soon as one link makes contact with the item, the next link unwinds: this is accomplished by having the pulley wheels at the tip of the tentacle smaller than those at its base. In such a way the tentacles can accommodate to the irregular shape of an object and ensure that it is grasped without damage. The task of accommodating to the shape of an object can also be achieved, albeit more crudely and with much less precision, by simply providing a robot with a large number of fingers. A seven-fingered robot has been designed to grasp irregular objects in this way, each of the fingers being equivalent in a way to the pins in the Omnigripper.

It is obvious that modern robotics is in a position to exploit a wide range of gripper options. Today it is no surprise to encounter such technological fertility in design and fabrication; but it is not always realized that great strides were made many years ago. Robotics technology is already drawing on a rich tradition, even though the age of practical robotics has so far lasted for only half a century. One relatively early project was carried out between 1973 and 1976 at the Fluid Technology Laboratory in Stockholm. It was supported by Swedish robot manufacturers ASEA, Ekstroms, Electrolux, Kaufeldt, Retab and Philips, and by the Swedish Board for Technical Development. As described by Lundstrom (1977), a collection was made of components actually handled by industrial robots of the day: around six hundred items were obtained, many of them geometrically very similar. No fewer than eleven component types were identified:

- Injection-moulded items using plastic (e.g. telephone covers, sanitary goods, push buttons)
- Die castings (motor frames, end plates)
- Packings (corrugated cardboard)
- Sand cores (for castings)
- Punched items (ball bearing rings, cartridges)
- Power-pressed items (ferrite cores, motor parts)
- Disk-shaped items (ceramic wafers)
- Sweetmeats (creams, rolls)
- Non-stable items (car parts in rubber, PVC disks)
- Reinforced items (glass fibre matting)
- Jigs (platens)

To this list were added a number of other components in regular use: work tools (spray guns, chisels, pliers), fasteners (screws, nuts, rivets) and assembled parts (integrated circuits, gear wheels, springs).

The choice of a suitable gripper for an application is crucial, since it comes at the important point where the robot interfaces with the workpiece. Choose an incorrect gripper and there could be very costly consequences: expensive machined items and assemblies could be damaged, and complicated production lines could be brought to a halt. A gripper may cost as much as a fifth of the total robot price – so it usually represents a considerable investment. In this context the particular advantages and disadvantages of individual gripper designs should be considered with care. A multifingered gripper may grasp an object too tightly; a gripper using a permanent magnet may attract metal swarf; a gripper relying on magnetism may not be efficient when a flat item is slightly bent (or if the magnetic properties of the workpiece are unsuitable); vacuum grippers using rubber suckers may be subject to corrosion; and so on and so forth . . . .

Fingered grippers, no doubt because of their humanlike character, have always attracted particular attention – and continue to do so. For example, the researchers Crossley and Umholtz (1977) studied the manipulative movements of the human hand, and later offered a design for a three-fingered gripper. A primary objective was a telechiric hand for manipulations in space observed by a television camera, so efforts were made to produce an anthropomorphic hand that could exploit the skills of the human operator. The result was an artificial hand with two fingers and an opposed thumb ('three fingers', in the jargon).

Another fingered gripper, the immensely versatile Utah/MIT dextrous hand, has invited much research and development. This complicated device is a 16-joint, four-finger manipulator principally applied in light assembly and repair tasks. The researchers Perlin, Demmel and Wright (1989), from the Robotics Research Laboratory at New York University, have explored a range of simulation methods designed to portray the various grasps of which the Utah/MIT hand is capable, and the specific tasks which the device may be expected to perform in practice. They saw a need to develop the scope of the hand in a practical direction ('The fact that the original designers of the hand at Utah had no specific end uses in mind and primarily developed it as a research tool for investigating control methods has further encouraged some of these more negative impressions of the dextrous hand'). Some difficulties with the hand in practical applications were indicated – for instance, it does not easily hold a screwdriver – and the possibility of design modifications was suggested.

Other recent research has focused on how flexible fingers can improve gripper sensitivity in such tasks as inserting electronic components in assemblies (Irwin, 1988); on how a model of the human 'prehensile function' can be used to develop a device-independent gripper controller (Liu et al, 1988); and on how a self-adaptable multifingered robot hand can be designed (Rakic, 1989). In this last case, there is again much emphasis on the anthropomorphic approach. It is emphasized that 'simplified grippers cannot provide reliable grasping of complex shapes and stable holding of these objects during the robot's arm motion'. As a result much effort has been devoted to developing hands 'with 3–5 fingers, each having 2–3 phalanges [segments]'. The hand considered by Rakic has four fingers with three phalanges each and a thumb; three servomotors, an economical number, are required to activate the hand. With this innovatory design, the hand can achieve a number of basic 'grasp modes' (hook, fist, pinch with thumb and one finger, pinch with thumb and two fingers, and lateral grasp with thumb), can achieve automatic finger position, is suitable for expert system control (see Chapter 4), and is small and lightweight. For these reasons it is hoped that 'this hand solution will find application with robot systems of the next generation'.

Grippers, of whatever sort, are true 'end effectors': they are generally sited at the extreme of a flexible or rigid structure, the robot arm. It is obvious that arms are a key element in robot design. They help to define the capabilities of the device, the possible applications that may be handled, and the extent to which the system is adaptable in circumstances that may not yet have been encountered. Some robot arms are relatively small articulated extensions, perhaps devoted to small assembly work; others are vast, robust structures able to extend 50 ft (15 m) or more and to handle prodigious weights. The manoeuvrability of the robot arm is generally regarded as a key feature in defining the competence of the system.

A typical robot arm has six rotary axes of motion, each driven by an actuator that in turn is controlled by a servo-valve; signals representing the changes in angular position and velocity of the axis are fed back to the control system by means of a resolver and a tachometer. So the machine intelligence is constantly made aware of the state of the overall system: it knows where the arm is, what it is doing, and what further actions are required to achieve the defined objectives. Like any sensible biological system, the robot uses information about the world and its own state to decide on a future course of action.

Just as hands may have two or more fingers, so robots may have one or more arms (though any number greater than two would be a rarity). Two-armed robots, as with the Vale devices already described, date back to the very origins of modern robotics, and there has been a constant supply of fresh two-arm designs ever since. More than twenty years ago the Sterling Detroit Company in the United States introduced the two-armed Robotarm to carry out the industrial functions of quenching and trimming in the steel industry, advancing the levels of automation that were necessary because it was becoming increasingly difficult to find men to perform such hot and unpleasant work. It was found that the original one-arm system encountered a number of problems, which were overcome when the robot sprouted another appendage. Following the design of two-arm systems it was possible to process castings two at a time in a second work cycle, the two arms precisely controlled to work in harmony.

When the die-casting machine opens, the gripper head on arm A advances between the die halves, captures the casting, travels axially and extracts the casting. Arm A then retracts the shot to a further position where the system confirms that the casting has been removed. Then the arm swivels the casting through 45 degrees and advances it for transfer to the dip quenching unit. Arm A then swivels to the start position while the casting is quenched. At the moment the die-casting machine opens, arm B captures the casting at the quench station, retracts it, swivels it through 45 degrees and advances the casting to the trim die. The withdrawal of arm B from the trim press die area indicates the beginning of the trim cycle.

The two-arm facility has found many applications in industry: it is easy to see why possession of only one arm may represent a severe functional handicap. But it is also clear that the provision of two arms rather than one can introduce many control problems. Robots with two arms should always know what they are both doing: it is important for the right hand to know what the left hand is up to!

Robot arms can be made of metal, plastic, carbon fibre and other materials. It has been found that composite substances – strong and light – are particularly good for robot applications. For instance the Locoman robot, developed by the Wolfson Industrial Unit in the UK to aid material forming, was designed with a three-dimensional pantographic arm made of carbon fibre reinforced rods (the system has also been demonstrated writing with a felt-tipped pen). Again, computer simulations have aided the design of robot arms in many ways. One example is the TUTSIM program that allows the generation of a graphical depiction of robot operation (Reynolds, 1986).

Other research work has focused on how two-arm robots can be induced to behave in a harmonious way (Grossman et al, 1985); how – in addressing the same problem – specific arm trajectories can be planned (Shin and Bien, 1989); and how specific control algorithms can be developed to supervise the performance of high-speed and high-precision robot manipulators (Zuren and Baosheng, 1988). In all such efforts an attempt is made, using high-level mathematical tools, to make robots behave in ways akin to those that human beings manage so effortlessly.

The Need to Travel

It is obviously useful for human beings and other animals to be able to move about in their environment (some plants manage it as well, but with less imagination). There are many problems associated with the task of movement, a fact that is only fully appreciated when humans try to design machines with this sort of ability. Problems of energy, balance, friction, limb co-ordination and navigation arise, to name a few. In biological evolution a host of special mechanisms have emerged to enable creatures to progress from A to B in a relatively painless and effective manner. Such mechanisms relate to aspects of anatomy, energy distribution and information processing. In coping well with the matter of mobility, human beings demonstrate that they are reasonably well-engineered physical systems. The task for the ambitious robotics engineer is to design artefacts that have anything like the same degree of competence.

As with artificial arms and hands, the manufactured leg has a lengthy history. There is a mention of an artificial leg in the *Rig-Veda*, an Indian account that dates to between about 1500 and 800 BC; and the Greek historian Herodotus mentions an artificial foot that Hegisistratus of Elis made for himself after lopping off part of his own foot to escape from the stocks. Artificial limbs are also encountered in the Talmud, the Nordic sagas and many other ancient texts. It may be expected that the rich could commission the best artificial legs, but at least the poor could lay claim to wooden ones: a Brueghel painting of 1568, now in the Louvre, Paris, shows beggars wearing wooden legs (was there a waiting list?). One famous leg, only partly wooden, was constructed for the first Marquis of Anglesey in 1815. This device, still in existence and sometime exhibited, has a steel knee joint and a wooden ankle joint, with cords from the knee controlling the motion of the ankle. Because it made a clapping sound when used it was dubbed a 'clapper' leg (other versions are known as 'American' legs). This leg is an obvious improvement on the wooden legs associated with pirates, and on the peg leg shown in a Roman mosaic at Lescar in the French Pyrenees.

A number of artificial legs were made by Ambroise Paré, already met, for the benefit of military amputees. Such devices typically included a movable knee and tarsal section of the foot, with a knee lock to provide rigid support at appropriate times. Artificial legs of this sort were obvious ancestors to the skilfully engineered prostheses of the modern age.

Various multilegged artefacts – walking tractors and mechanical horses – have been employed in agriculture for many decades, though with mixed success. It is reported that the Soviet Union has employed stepping excavators to clear forests, though such machines seem to have been experimental devices rather than effective working sytems. Two Soviet researchers, Artobolevskii and Kobrinskii (1977), have provided details of a four-legged mechanical machine weighing 1½ tons (1½ tonnes) and powered by a 90-horsepower automobile engine. Such designs have made little impact on the technological community.

General Electric has experimented with walking devices, as have the University of Wisconsin (to study problems of stability and control), medical workers trying to provide amputees with mobility, and other researchers with a particular focus of interest. For more than thirty years efforts have been made to develop the theory of adaptive walking machines. Here the work of R. B. McGhee and his colleagues (Orin, McGhee and Jaswa, 1976; McGhee, 1977; and McGhee and Iswandhi, 1979) has been of particular importance. A hexapod robot that has been operating at the Paris VII University since November 1980 is described by the researchers Kessis, Rambant and Penne (1982). Again such a device can be regarded as having educational and entertainment value rather than being of much practical use. It is of course easy to speculate on how mobile robots might be employed in practical activity. There are obvious industrial uses in the factory environment (mobile systems for use in factories have already been discussed) and elsewhere in circumstances in which wheels or tracks would be unsuitable. Legged robots could have advantages in such applications as Arctic transport, mining, agriculture, forestry, firefighting, bomb disposal, ocean floor surveys and planetary exploration. There are plenty of reasons for conducting research into mobile robots.

As with other types of robotic features, mobility can be provided in many different ways. Wheels and tracks have been mentioned, but these tend not to be favoured. Although wheels can be used on autonomous guided vehicles in the factory and in other environments, a flat surface is essential: the provision of wheels implies that 'roads' of some sort are available. Tracks can be employed to offer mobility over rough terrain, but they tend to damage delicate surfaces and are less accommodating to certain types of environmental irregularities than are legs; a legged system can, in principle, step aside or step over, being flexible enough in operation not to have to continue in a more or less direct line. The advantages of legged system, do not stop with their functionality. Legs are clearly suited to our concept of an anthropomorphic robot: a mobile artefact without legs may seem to us less of a robot.

Just as the number of robot fingers can vary from one system to another, so can the number of legs. Robot have been designed with only one leg; such hopping devices, with balance maintained by gyroscopes, are remarkably effective – though they tend to be unstable when they are not moving. There are also many different robot designs for machines with two, four, six or eight legs (and some mobile robots rely on snakelike movements). The different types of legged systems all have advantages and disadvantages, but some of the problems that they face are common to them all. For example, all legged systems have to maintain adequate stability, both when at rest and in motion; facilities for steering are essential if the robot is to travel with any useful or intelligent purpose; and a method of powering the active legs has to be included. Where such requirements are not met, the robot will be at best inefficient, at worst a hazardous threat to human beings in the environment. It is easy to see that a robot might fall when attempting to climb stairs, that it might collide with objects if it is not able to navigate and steer, and that any on-board power supply will be quickly depleted if its resources are wasted on powering a poor design (leg joints have to be activated, and power can be wasted in stopping and starting the relevant motions).

One-legged ('hopping') robots have an uncomplicated gait but they need a very sophisticated system to maintain stability: a quadruped that always has three legs on the floor will never fall over, but a hopper has no such privilege. In a typical design the solitary leg slides in a cylinder that also carries a strong spring:

the sliding movement can be powered down vigorously to raise the mass of the system, so causing the device to hop. At the same time rotary power can be applied to control the direction of movement and to contribute to the stability of the robot. The device may have a lead linking it to an electrical power source, but such a provision is an obvious limitation on its mobility. Energy losses would occur when the device strikes the floor and when the spring extends to hit a stop positioned in the cylinder.

Hoppers, therefore, are of limited application. If they are to carry any load for transport purposes, then the stability problems are bound to increase, as indeed they do if an on-board power pack is included or if the robot is required to travel over rough terrain.

The design of robots with two legs faces a range of characteristic problems. Again there are difficulties associated with balance, navigation and power; but the situation is complicated by the requirement that two complicated legs are controlled to cooperate at all times. Biological systems are able to collect sensory data to aid the task of maintaining balance: if humans or animals see a step or an incline in front of them they adjust their performance accordingly. But it may be difficult to provide a robot with a vision capability (see Chapter 4), in which case it is limited – like most wheeled vehicles – to travelling on smooth and unproblematic surfaces.

One famous two-legged robot has been constructed by I. Kato and his colleagues at Waseda University in Japan. This device, the original Wabot system, has large, platelike feet, an ankle and pelvis, and is controlled by a computer. The design includes eleven rotary joints; two yaw joints on the pelvis allow one leg to stride in a different direction to that of the foot already on the floor, so allowing the robot to walk round corners. The joints are individually driven by a hydraulic actuator, with joint displacement information continuously fed back to the computer for control purposes (an impressive descendant of Wabot will be described in Chapter 5).

Much work has been carried out to research aspects of human walking, skipping, running and other modes of activity. This, as with the design of artificial hands, has provided many clues for the design of two-legged robots. It has also helped with the design of prosthetic legs for amputees, and of powered exoskeletons for people with paralysed limbs. For example, scientists at the University of Wisconsin in the USA have built a powered walking machine for paraplegics. This uses a ⅓-horsepower motor, artificial hip and knee joints worked by hydraulic actuators, and a digital computer for control purposes. The system is taught by a normal person who climbs inside the machine and makes the required movements: in this way the exoskeleton can be programmed to sit, stand, walk, avoid obstacles and climb stairs.

Human beings are regarded as having two gaits: walking, when at least one foot is always in contact with the ground; and running, when both feet can be out of touch with the ground at the same time. In one sense this is an extremely unstable situation. It is only the movement that maintains stability: freeze it in time, and the system would at once fall over. By contrast, the horse has four gaits: walk, trot, canter and gallop. Such considerations have led to the development of a theory of generalized gaits, usefully applied in the design of mobile artefacts. All possible gaits have been enumerated, with the focus on stability features, though many other factors too are important in choosing a design option.

Many four-legged designs are relatively unstable, unless they have wide feet or roll the body sideways, as does the two-legged Wabot. One particular Japanese system achieves static stability by implementing a four-legged crawl, as do some of the artificial crawling babies described in Chapter 2. Here three feet are always in contact with the surface, and the centre of gravity of the machine is always within the area of the three feet.

Four-legged systems also have characteristic problems in connection with navigation and steering – many systems have to stop before a successful change of direction can be accomplished. Some systems include a fifth leg to aid turning: the leg is pushed down and the body then rotates until it faces the required direction. It is difficult to organize non-stop turning without a computer facility or some other device to enable leg movement arcs of the required size to be made. Such steering problems also apply to six and eight-legged robots – which, however, are able to guarantee permanent stability when static.

When a robot has six or eight legs, like an insect or a spider, three legs can be used at all times to form a stable triangle. The movement of the other legs is arranged to ensure that the centre of gravity never strays outside a stable base, so the robot never stands any risk of falling over – always assuming, of course, that it never attempts activities well outside its scope (like trying to climb a vertical wall, if it is not a wall-climbing robot). Here, as with simpler systems, data can be collected from the many joints and fed to a computer which is thus equipped to set up subsequent leg movements. Again it is very useful if the robot has sensory equipment to aid navigation, the avoidance of obstacles, and autonomous initiatives to counter unprecedented hazards.

It is obvious that the more legs a robot has the more complicated will be the associated control facilities. The possible number of operational sequences becomes progressively greater with the number of legs. We have seen that hoppers have only one possible sequence, and bipeds two. But even a tripod has as many as six sequences (the three-legged hop, two legs moving at once, and one leg moving at a time); while a quadruped has twenty-six possible sequences of leg action. Clearly a facility for motion provides a robot with additional versatility (a further three degrees of freedom, in the jargon); but there are many complications in providing an artefact with mobility: Such difficulties have been addressed in many ways, but modern mobile artefacts – whether legged or not – are primitive creatures, and the research continues.

Efforts are continually being made to learn from the mobility features of biological systems. The researchers Fichter and Fichter (1988) have explored the 'kinematic' characteristics of insects and spiders with a view to applying the findings to the design of legged robots. Leg designs have been examined in connection with segment length relationships and the specific mobilities of the body-leg system. It is of interest that, for example, insect legs are more specialized than those of spiders, and that in consequence insects require position control for fewer joints. Other research (Kimura et al, 1988) has focused specifically on improving the walking performance of artificial quadrupeds. Here the relationships between stability, maximum speed and energy consumption were formulated with specific attention to the characteristics for locomotion (gait, length of leg, period, stride, joint angle and so on). Experiments were carried out with the quadruped robot Collie–2, a functional computer-controlled system.

Other research has focused on how legged robots can be designed to interact efficiently with the rough terrains to be found in agriculture, mining and construction, and on other planets (Waldron et al, 1988). And some work has investigated the requirements for legged robots in specific environments. Nagy and Wittaker (1989), for example, consider a novel six-legged robot design in the context of planetary exploration. In this case efforts have been made to develop a sophisticated motion control scheme, the robot design lending itself well to predictable and reliable performance in an alien environment.

The provision of artificial legs for robot systems can be regarded as a largely predictable evolutionary development. One of the main themes of this book is that robots with anthropomorphic features will develop in the direction of *Homo sapiens*. Biological evolution, by dint of trial and error, has generated countless species that are well adapted to their specific environments. Human beings have an interest in designing and fabricating artefacts that can behave intelligently in human environments: we can speculate (see Chapter 6) on how the humanoid robot is likely to emerge as colleague and counsellor, as friend and lover. In such circumstances it should not be a surprise to witness the robot emerging as a humanoid creature. Timescales are debatable, but the direction of development – for at least one species of artefact – seems unambiguous.

Motive Forces

Every robot requires an effective actuating system to provide it with useful muscle power. In particular, it needs devices that can convert energy from one form to another. Robots are active in the world; and they typically derive their mechanical power from electrical, hydraulic or pneumatic sources or some combination of these.

The typical DC (direct current) motor is a common source of motive power for modern robots. Robot designers can draw on a wide range of different motors; and they are easy to control and power in the typical industrial environment. The high speed of the motor is geared down to activate a robot arm and any other mobile robot element at the required speed; elaborate gear reduction units have been designed for this purpose. DC motors also aid the control functions in various ways. For example, when a gripper firmly grasps an object or when an end effector reaches the limit of its travel, there is alteration in the amount of current drawn by the motor; this change can be detected by the control system and fed to the computer as important information about what is happening during the process being carried out. Today a growing range of motor options – conventional DC, disk, 'pancake', brushless and so on – are available to robot designers. Stepping motors are of particular importance.

The stepping motor is able to provide the precise programmable position control that the robot needs. Unlike the typical DC motor, the stepping motor rotates in small steps, each a known amount. The control computer, carrying information about the geometry of the robot and the gearing of the motors, counts the number of steps to determine the position of the various robot elements activated by the stepper motors.

Hydraulic power can be used to generate mechanical force by pressuring a liquid (the word 'hydraulics' derives from the name of a Greek automaton, the *hydraulos*, a musical organ driven by water power). The hydraulic approach can provide a flexible linkage: unlike a mechanical link, a hydraulic pipe can loop round corners and obstacles; and it is also highly efficient. Large industrial robots are typically driven by hydraulic power, as are the massive excavators used to dig up roads. Hydraulic systems can deliver immense power, but they tend to be expensive and there are obvious problems if fluid leakage occurs.

Pneumatic machines are driven by air pressure (the Greek *pneuma* means breath or air), or sometimes by the pressure of inert or hot gases. The earliest systems used air (see McCloy and Martin, 1980), but today there are many other options. Pneumatic systems cannot deliver anything like as much power as hydraulic ones, but engineers can design for much lighter units since gases weigh considerably less than liquids. This means that pneumatic systems are likely to be cheaper, but there can be problems where precise control is needed. Gases are highly compressible – which may make the movement of an end effector slightly elastic (an advantage in some applications, a disadvantage in others). Moreover, compressed air systems can be expensive and noisy.

It is also possible to supply robots with electric power, a precise and inexpensive way of activating a robot arm or other moving parts. Electric actuators are quiet and clean, and are likely to offer various design advantages over cumbersome piping; but the torque/weight ratio tends to be low, and the possibility of arcing and electric shocks can represent safety hazards. It has been seen that electric power can be provided via a range of DC motors, stepping motors and so on; and that such options are widespread in modern robot design.

Robot actuators are essentially required to convert hydraulic, pneumatic or electric power to mechanical power. The advantages and disadvantages of the various types need to be considered with care by robot engineers. It should be remembered that robot design should be based not only on cost/performance factors but also on the constraints imposed by important environmental considerations, and on the needs of human beings.

This chapter has focused on some of the key features of modern robotics. In particular, attention has been paid to the structural features of characteristic working systems, and to robot elements that have real anthropomorphic significance, such as artificial hands, fingers, arms and legs. Some working systems have been cited, but the items included here are very far from an exhaustive compilation. In addition to the functional systems mentioned in this chapter there are also the six-legged Odex I walking machine (from Odetics Inc), the three-fingered Salisbury hand, the Whittaker Workhorse (remotely controlled to work inside the highly contaminated Three Mile Island nuclear plant in the USA), and the articulated ROSA system designed by Westinghouse to examine and repair tubes in nuclear power plants. Such systems draw on aspects of robot anatomy to perform tasks in the real world.

It has been emphasized that automata can encompass many different types of systems: for example, telechirs and exoskeletons (highlighted by Thring, 1983) are important cousins of anthropomorphic robots. This chapter, however, has focused on artefacts that are evolving, in diverse ways, in the direction of *Homo sapiens*. Such systems offer the most dramatic possibilities for the future.

It should also be pointed out that this chapter has only briefly mentioned some features of robotics that are essential in modern design. It has alluded to such topics as sensory capabilities, programming, control and the possibility of intelligent robot behaviour. Such matters, vital for the evolution of sophisticated robots, are considered in Chapter 4. Modern robots are not only evolving physical features that can be compared with those that have developed in biological systems; they are also evolving a spectrum of faculties that exhibit mental characteristics. Modern computer technologies are making it possible for robots to acquire minds – with all the implications that such a development implies. Emotion, free will, creativity and mental disease are not necessarily beyond the scope of artefacts. The time will come when it will seem unreasonable to deny that the robot – by dint of a progressive accumulation of physical and information-processing elements – has become a person.

4 Intelligent Machines

The concept of autonomy is important to the intelligent robot: it has to be able to 'act on its own'. For many centuries there have been thousands of different types of machines, structured in particular ways to perform characteristic tasks; but few of them have been in any sense 'self-moving' or 'self-controlling'. A key feature of intelligence is that the system has a degree of independence from the environment, the capacity – at best – to take decisions, in changing circumstances, in furtherance of a goal.

This means that intelligent systems have to be interested in information – about themselves and about their world. They have to know what is happening so that they can take compensatory action. In short, they have to be able to collect data and then process it for particular purposes: the intelligent system, required to work effectively in the real world, needs senses and a brain. And it needs other elements as well if it is to be a true cybernetic system: a system capable of adaptation in pursuit of objectives.

We can debate what is meant by autonomy and functional independence (see Chapter 7), so broadening our insights into the nature of intelligent systems, including human beings. We can note, for instance, the importance of programming to the most sophisticated robots; and consider whether the provision of programs is compatible with functional independence. Put another way – can a programmed system behave 'freely'? Are human beings programmed? If so, what does this say about *their* autonomy? If not, how are they able to carry out operations in the real world that require the close observance of orderly sequences and procedures? Such questions, important to much of the rest of this book, help to throw light both on the nature of *Homo sapiens* and on what may be expected from intelligent robots.

Controlling the System

All functional systems – for instance plants, insects, horses, cars, robots – have to be structured for coherent control; otherwise their constituent elements will fall apart or work in mutually antagonistic ways. The study of system architecture is largely a study of how the system is controlled. How does this part link to that? How does a change here affect the situation there? How does a physical input stimulate internal activities to generate a physical output? The robot, a useful paradigm for many types of high-level systems, particularly those in the biological world, has a tightly structured architecture so that its various operations can be performed without ambiguity to accomplish goals. There are many ways in which robot control can be realized in real-world system. The most sophisticated robots have computer brains and a wide range of sensors (for sight, touch, hearing and so on,) but many early robots did not enjoy the versatility offered by modern electronic computers.

The most primitive types of robot control can be achieved simply by means of physical set-up: cams, stops and levers may be positioned to ensure that the robot runs through a number of steps which, taken together, define the required task. This type of programming allows for some flexibility: a stepping drum, for example, can be reprogrammed to define a new task, though such reprogramming generally involves human intervention. Many early robots were programmed in this straightforward physical way to enable the system to function reliably and predictably in an industrial environment. Early Prab robot systems, for instance, contained stepping drums with reprogrammable cams that operated electrical switches which in turn supervised hydraulic actuators to control the robot arm. In this way such operations as die castings, forging, plastics operations and material handling could be carried out with precision and in a totally reliable fashion.

Other early control systems relied upon the exploitation of pneumatic possibilities. In the 1970s the Auto-Place SPM–4 control unit used air and electrical signals to control the movements of a robot arm; and improved versions of the SPM units were in wide use in the 1980s. The early air-operated devices comprised a step-sequencing module and an air-cylinder power module. Poppet indicators for the various sequencing steps were set across the top of the sequencing module, with similar indicators on the power module allocated for the various operational steps. The sequencing module was used to send to the power module information about which cylinder action should take place, whereupon the power module initiated the cylinder action, sensed completion and informed the sequencing module. The sequencing module was then able to advance to the next step and initiate a new operation in the procedure. The Auto-Place control unit could initiate up to twenty-four consecutive operations in this way, so supervising the robot to perform a relatively complicated task.

It was also possible, using limit valves and other external devices, to provide conditional control over the sequence of operations. The system was able to accept electrical signals which could then be used to operate solenoid valves able to control the air flow; conversely, air signals could be converted to electrical signals via the appropriate system elements. All the relevant connections were made in a defined sequence to perform the required task. The system could be reprogrammed, but only in a tedious and time-consuming way.

A wide range of complex industrial and other tasks can be performed using simple physical set-up, pneumatic logic or electrical sequencing methods. Such approaches do not, however, allow for complex decision-making or the performance of highly complicated tasks required hundreds of operations, some of which may be conditional upon local environmental factors. Such flexibility of response, such intelligence, can only be accomplished by computer control: all modern sophisticated robots have computer control facilities, often with intelligent semiconductor devices encapsulated in their various parts (individual limb joints may be assigned a degree of local intelligence, as may individual sight or touch sensors).

Robotic control theory largely relates to the movement and response of mechanical joints in carrying out tasks. Again it is worth emphasizing that robots are active systems in the real world. They collect information, not for their own amusement, but so that they can function effectively in a particular application or range of applications. Typically, information about joint position may be fed to a local microprocessor which may then generate further control information or provide input to a master computer which will then generate its own control outputs. It is also clear that robot intelligence will increasingly require an extensive database, an effective corpus of knowledge about robot features and the wider world: true intelligence cannot function without memory, images, working models that can be constructed and adapted in a

12 Omni 2000 robot for home services and entertainment

13 Dingbot robot, a toy system ('crazy and fun-loving') that acts capriciously

14 & 15 FEDMAN industrial robot (*above*) and
ASEA 2000 industrial robot (*below*), both able to perform in a wide range of manufacturing applications

FEDMAN – M

素材置場
ASEA

16 Sarcos Dextrous Arm used in conjunction with a force reflective master to form the Sarcos Dextrous Teleoperation System (DTS), enabling a human operator to carry out tasks in a hazardous environment using teleoperated control

17 (*opposite*) NEATER (Nuclear Engineered Advanced Telerobot) for automation in the nuclear industry

STÄUBLI
UNIMATION
PUMA 700
STÄUBLI
UNIMATION

18 Industrial robot, used for welding and other applications

19 Industrial robot with camera vision facility

20 Industrial robots used for welding on Ford Fiesta manufacturing line

mind. This suggests that the truly autonomous robot will need a high level of artificial intelligence.

The requirement for more complicated robot operations has encouraged the development of more sophisticated control systems. Use may be made of analogue voltages and dedicated semiconductor logic for particular control functions, with the various facilities selected 'off-the-shelf' or designed in a tailor-made fashion. There are thousands of options to choose from: a number of robots may be controlled by an external mainframe computer or minicompter system, or arrays of microprocessors may be configured for particular purposes. A special-purpose computer may offer inadequate flexibility for future expansion; and this or that off-the-shelf system may be only partially adequate, strong in some features and weak in others (it may not, for example, be sufficiently robust in a particular industrial environment). The important point is that all the relevant factors should be taken into account: current requirements have to be met, and there should also be scope for the fresh requirements that are likely to emerge in the future.

Robot controllers, today usually based on semiconductor logic, are becoming ever more powerful and ever smaller. In the 1970s it first became clear that an increasing amount of robot intelligence would be provided by microprocessors, rather than by a bulky and expensive mainframe sited away from the robot, sometimes in a separate room away from industrial hazards. One early microcomputer system was provided by the Metal Castings Company at Worcester in England for a Series 2000 Unimate robot. Despite the compact nature of the microcomputer controller it was still necessary to site it some distance away from the robot: an underground 'umbilical cord' was provided as a linking facility. In this application the robot was used to serve a 600 ton (600 tonne) Triulzi 600T die-casting machine used to produce aluminium components for cookers, cars, refrigerators and office equipment. Here the special advantage of the microprocessor was that the die-casting work could be quickly reprogrammed to enable different parts to be manufactured. The micro specified the required sequence of operations, and also checked the safety of the overall operation: like many modern systems, the semiconductor intelligence was programmed to ensure that guards were closed and that there were no obstructions in the vicinity (some clever modern robot systems can detect whether there is a human being within a 'no-entry' envelope). Only when all the necessary environmental checks had been completed could the operation, potentially hazardous to anyone in close proximity, be initiated. Microcomputers are today widely used to monitor the performance of robot sensors and to teach the robot to perform required tasks. The various advantages of micro systems are clear: they enhance sensory capabilities and improve machine intelligence, can be sited in small spaces (including many 'on-board' sites) and are relatively cheap.

It is important to emphasize the size advantages offered by modern semiconductor-based logic systems. A robot linked to its brain via a lengthy umbilical cord is limited in many ways: in particular, it lacks mobility – which may not be important on an industrial production line, but which is likely seriously to hamper the performance of robots in the future. How much better if a robot, like any self-respecting biological specimen, can carry its brain around with it (linked also to an internal power source), allowing freedom of movement in the environment – a true condition for proper autonomy.

The compact size of modern semiconductor logic also enables artificial intelligence to be distributed around the robot architecture: in such a way the robot brain may be co-extensive with the robot body, an interesting evolutionary arrangement. There are obvious advantages in having many information-processing tasks carried out on a 'local' basis, near to the anatomical site where the control signals are required. In particular, such a facility enables the robot to respond quickly when a new action needs to be taken. It may not take long for a robot hand to detect an environmental threat and to send a signal to a robot 'head' some distance away and then for the head to activate a hand withdrawal or other compensatory action – but why bother if the hand can be equipped with its own intelligent logic and sort the matter out for itself? This suggests the provision of a truly distributed intelligence to enhance the robot's speed of response and to optimize its behaviour in changing environmental circumstances. At the same time, it should be remembered, there are various problems in assigning local autonomy to what is intended to be an integrated system devoted to a particular task or range of activities.

It is also worth remembering that there is an important sense in which the human brain works as a distributed information-processing system. Thus Garrett (1978) noted that 'the human brain is organised as a distributed processing system' – an arrangement that is increasingly common, not only with robot systems, in the world of artificial intelligence. Biological and artificial information-processing systems accept information – from both outside and within themselves – store it, organize it and process it; and use pre-existing programs to generate fresh programs and to yield information outputs. Different parts of the mammalian brain are allocated to different senses and to other specific functions, much in the way that different microprocessors might be dedicated to different robot subsystems. But it is obvious that only minimal information processing is carried out, in most biological creatures, away from the main brain centre: the human brain may well function on a distributed basis, but this is more an operational than a geographical distribution (few of us have bits of brain in our hands and feet, though incoming sensory information is shaped and coded in various ways).

The design of robot brains will continue to be influenced by what we know of biological information-processing systems; but at the same time we are not constrained by what natural evolution has managed to achieve. There are many ways in which we may be able to do better than nature. Intelligence, rather than trial and error alone, will continue to play its role in the evolution of robot capabilities: in fact both human and machine intelligence are being used to shape the course of robot evolution.

Aspects of Computer Control

It has been shown that the evolution of control systems for robots moved from simple physical set-up to simple pneumatic systems and finally to semiconductor logic controllers. The earliest semiconductor devices were not very intelligent: at best they resembled the simple semiconductor devices used to control petrol pumps. In any event, all the early controllers lacked flexibility: it was tedious, and sometimes difficult, to reprogram drum cams, pneumatic-logic pipes, electrical connections and the like; and semiconductor circuits often allowed no reprogramming options (a new task may involve throwing away the circuit and fitting another one). True programmable flexibility arrived with computer control.

Today both microcomputers and minicomputers are used to control robotic systems in fixed and flexible applications; sometimes a large mainframe system may be used to control a robot installation, but this is increasingly uncommon – there are many advantages in having a local dedicated intelligence. Modern computer control facilities allow effective monitoring of sensor 'experience'; offer a range of self-diagnostic options; and provide access to libraries of programs, an alternative to reprogramming an existing system.

The memory facilities, usually based in semiconductors, provide both primary-store and secondary-store systems. Primary (sometimes called 'buffer') stores carry transient information that may be required briefly, allowing rapid local processing that quickly yields an output and then just as quickly moves on to other matters. Secondary ('backing') stores hold information for a lengthier period, sometimes on a permanent basis. In this case the information is of permanent use to the robot: if it forgets secondary information it may be in real trouble! The robot can draw on both the primary and secondary stores to carry out processing tasks that are essential to the task in hand. The computer may carry out parallel processing ('multiprocessing'); for example, in simultaneous control of the different joints in a robot arm. Or it may encourage dedicated microprocessors to get on with their own special tasks. Many modern robot control systems mix an element of autocracy with an element of democratic delegation.

The more sophisticated the robot, the greater the range of tasks that the control system is required to supervise. Even in the simplest robot applications the control system is required to supervise the position, movement and performance of the end effector; and such requirements may involve the monitoring and control of arm acceleration, deceleration, amount of travel and attendant forces. It is also often necessary to control the motion of the arm from point to point along a route: it may not be sufficient simply to ensure that the robot arm gets from A to B by any route it chooses. The computer also needs to be able to control all phases of an operation, supervising the robot through all the constituent steps of a process. This task may be rendered much more complicated when external equipment is involved in the activity. A robot may, for instance, be cooperating with a machine tool, a metal press, a forging device or another robot; if a companion machine goes wrong, or behaves in an intelligent but unprecedented way, then the robot has to be able to respond appropriately. This requirement may represent a further burden on a hard-working control computer.

Robots equipped with senses have access to a constant flow of data from the real world, and such data has to be collected and processed. It may be necessary, for example, for the computer to compare data derived from touch sensors with data derived from vision sensors (robot 'eyes'), in order to ascertain the true shape of an object which the robot has encountered. And there is always the matter of expectation. Did the robot expect to encounter the object? Does it belong in the defined task? Perhaps the object was one of three, or a dozen, or a hundred, that the robot 'knew' might turn up; or perhaps the new arrival is totally unexpected. What is the computer to do in every case? Is it equipped to cope? If not, what are the emergency routines? Is there a fail-safe provision in the event of robot 'experience' that cannot be interpreted?

Computer control units often come with encapsulated programs set in semiconductor circuits; or the computer – and so the robot – may be programmed by the operator using a typical computer keyboard. This latter approach is effective when the task to be performed is relatively simple; but a complex operation, requiring a complex program, could only be manually programmed in a very time-consuming way. Where a complicated task is to be carried out it is conventional to program the robot by showing it what to do: here the robot monitors a human performance and remembers all the necessary steps. 'Teaching-by-showing' is ideally suited to the training of robots in such application as paint spraying and component cleaning: the human operator moves the robot arm through the sequence of operations that are required for the full performance of the task. There are various ways in which this can be done.

With lead-through programming, use is made of a special teaching aid that can be attached to the robot arm; alternatively a special teaching arm can be employed for the purpose. The operator sets the system controller, defining various parameters including the sampling rate at which particular items of key information are to be recorded in the memory. He then operates an interlock which allows him to assume manual control of the robot arm, whereupon he conducts the robot through all the stages of the task in question. Special provision has to be included to enable the operator to perform the physical task of moving the robot arm through the work cycle: if the robot is not a small one it may be necessary to counterbalance the arm and to reduce any sustained hydraulic pressure so that the operator does not have to work against the drive system of the machine.

In a typical training sequence the operator would actually perform the task – for example, spray-paint a component on a conveyor. All the details of the operation would be remembered by the robot, including how and when to switch the spray gun on and off. One advantage of leading the robot arm through the desired sequence is that it is impossible, in this way, to program the system to perform tasks that are beyond its capacity.

An alternative teaching approach is walk-through programming, where use is made of a remote-control teach pendant to drive the robot arm to different positions which it then remembers. A simple pendant involves driving each joint of the robot arm to a particular position and then pressing a switch or button so that the position information is stored. More sophisticated pendants offer a range of useful facilities, including provision for time delays, manual or playback speed control, and emergency 'panic' stop facilities. In some high-level pendants there is provision for moving the robot arm in a defined fashion, simply by activating a single pendant function, even though the requirement may involve the simultaneous activation of several joints. Once the robot arm has been effectively programmed ('taught'), it can be activated to move through the sequence, either continuously or step by step, to check that it knows what it is expected to do. Once the operator is satisfied that the robot is well trained it can be let loose on the actual production task.

When the operator chooses to 'key in' a program, much in the way that a home computer or larger system would be programmed, he is involved in what is called textual programming: he proceeds much in the way he would when programming any sort of system for any sort of task. Many programming languages have been developed for the specific programming of robots, usually in an industrial context (some observers have said that increased coordination between the various programming options would be very helpful). The available textual languages include VAL (Unimation), AML (for IBM's assembly robots), SIGLA (for Olivetti's Sigma robots), RAIL (Automatix), TEACH (Bendix), INDA (Philips) and WAVE (for research at MIT). A typical textual language may begin from an awareness of the features of a particular robot range, and then construct around this awareness a syntax that will fully describe how the robot might be expected to move in all its possible applications. An alternative approach is to start with a more conventional language – one that already has many features that the user can exploit – and to adapt this for use with robots.

Ideally a textual language used in the context of robot applications will have facilities that satisfy a large number of requirements. For example, it should satisfy the day-to-day operator, who may be required to reprogram the system at frequent intervals; maintenance personnel, mostly interested in self-diagnostic routines, system reliability and other related matters; and applications programmers, who may have an interest in broadening the scope of the robot facilities far beyond the typical everyday operations. While it may be difficult to satisfy all these users, there are obvious advantages in employing the textual programming approach to sophisticated robot systems, particularly where a vast robot arm cannot be physically manipulated by its human teacher or when a very complex application cannot be taught using available teach pendants. It is likely that, at least in the near future, robot programming will be achieved by a pragmatic mix of textual and teach methods.

The programmed instructions, however accomplished, can only be implemented via actual equipment, the hardware configuration that in part defines the scope and character of the robot system. The central processing unit (CPU) that carries out the necessary computations is invariably structured out of semiconductor material 'doped' with certain chemical impurities: in some relatively simple CPUs there may be only one or two microprocessors dedicated to the required logical and arithmetic tasks. Micros carry transistor-type components and other electronic elements such as resistors, diodes and capacitors. They can function satisfactorily within a wide temperature range (some are found under car bonnets), consume little power (so dissipating little heat) and are relatively cheap. Microprocessors typically link to memory devices: most commonly, ROMs (read-only memories) and RAMs (random access memories).

Other items of robot-associated hardware include buses (channels for data from one component to another), power supplies (sometimes on-board, sometimes from an external source), inputs and outputs (allowing, amongst other things, the collection of data from robot sensors), and such teaching aids as keyboards, pendants and visual display units (small displays – using light-emitting diodes or liquid crystal displays – may be incorporated in the robot structure to signal what is happening in particular regions of the robot anatomy). Such pieces of equipment complement the various items mentioned in Chapter 3, such as DC motors, hydraulic power sources and the various pneumatic options.

The provision of computer control has made it possible for robots to develop as intelligent systems (considered further in Chapter 6). The gradual emergence of robot intelligence significantly expands their claims to a humanoid status: high intelligence, scarcely approached in most computer-based artefacts, is a suitably anthropomorphic phenomenon. Once robots have acquired a spectrum of mental powers their existence in the midst of human beings will come to have many implications. It is useful to emphasize that robots will be aided in their development of intelligence by a growing spectrum of artificial sensors, devices that will increasingly enable the growing family of humanoid artefacts to learn about themselves and about the complicated and ever-shifting world that they share with human beings.

The Need to Know – Senses for Robots

It is quite possible to program a robot to perform a wide range of tasks without it having any sensory awareness. Indeed all the early robots – the pick-and-place units and their cousins – were expected to function in this way: there was no requirement for robots to be 'aware' of changes in their environment, no need to provide them with any capacity for independent choice when their immediate circumstances changed. Such first-generation systems were useful, predictable and very limited. Designers were soon looking to a new generation of robotic devices that would be able to monitor their world and to act accordingly. It was inevitable that robots would begin to evolve not only brains but a spectrum of senses able to offer the information without which brains are massively constrained.

Again it has proved an easy matter to look at biological species in order to obtain quick insights into the sorts of senses that might be designed for artefacts. We are all acquainted with the five senses of human beings (we will leave the sixth sense to the optimists), and many efforts have been made to duplicate these in computer-based robots and in some other types of machine. But robots will not necessarily be limited to the five senses that characterize *Homo sapiens*. It can already be seen that ultrasonic ranging, well employed by bats and dolphins, is being employed by robots; magnetic effects, thought to be involved in some types of bird navigation, may come to be exploited in robot sensory equipment; and parts of the sound and electromagnetic spectra not accessible to any known biological species may come to be used by robot systems. We can speculate on how robots might use doppler microwave modules, exploiting the frequency shift in reflected electromagnetic waves, in long-distance sensing; and how familiar sensory strategies – vision, touch and so on – might evolve dramatic new forms.

It will be necessary to design control systems able to use increasingly sophisticated sensory inputs. This will assist robots designed to act in a goal-seeking manner: incoming sensory information will be compared with the existing robot knowledge before decisions are taken on an appropriate procedure, so duplicating the strategy of any human being trying to realize an objective. Sophisticated robots will increasingly be able to develop a world model that can be changed by new data collected by the senses: thus a robot will be able to learn from experience how to shape its perspective on reality.

The sensory equipment of a robot serves as an interface between robots (and other artefacts) and their environment (of which human beings are a part). Robot sensors should properly be regarded as a means of aiding the evolution of robot intelligence. In fact AI (artificial intelligence) facilities are increasingly being designed into sensor systems, so sensors have a double link with intelligence: they are evolving a useful local intelligence related to the immediate processing of sensory data (this parallels what happens in the sense organs of many biological species); and they are contributing to the intelligence of the overall system as it functions in an integrated way. Once again, a historical thread has contributed to the modern robotic weave: as with automata, the practical sensor has a long tradition. Artefacts in ancient China and India were sensitive to a range of environmental factors such as water pressure and air temperature. Centuries later, Galileo's thermoscope – a glass bulb connected to a glass tube dipped into a reservoir of coloured liquid – was designed to detect temperature and atmospheric pressure; in 1612 he constructed a tube linked to an alcohol-containing device, so inventing one of the earliest thermometers (Gabriel Fahrenheit was later to replace the alcohol with mercury). Soon an immense range of sensors were being conceived and built, developing the great family of artefacts that in the twentieth century would come to contribute to robot anatomies.

In recent years it has become increasingly clear that real-time computer systems (i.e. systems responding in time with actual events) could and should be developed to exploit information collected and supplied by artificial sensors. In an article called 'Artificial intelligence: sensors need it', Stephen McClelland (1987) argues that 'the next leap of faith for artificial intelligence will be to implement fully-operative real-time control systems, with the processing of sensory information immediately followed by analysis, followed by the proposal of decisions, followed by the execution of those decisions'. What is being suggested here is the increased autonomy of real-time control systems – which means in robot terms that machines will be able to sense what is going on and to act accordingly. The copious collection of sensory information and its subsequent processing will require massive computer power, as indeed human beings possess; but it should not be thought that this requirement will service as a permanent brake on robot evolution. Robot senses are in place and evolutionary improvements are being introduced month by month. Moreover, what robots are doing is already a model of what happens in the mammalian brain: information is taken in via the senses, initially processed at a rudimentary level, and then fed to the higher cerebral centres for appropriate interpretation and storage. This results in an extensive enlargement of the knowledge base, and the progressive emergence of new programs to provide enhanced behavioural flexibility.

It is also useful to remember that robot senses are not always concerned with collecting data from the outside world. It has already been seen that information about joint position, arm mobility and so forth can be usefully fed to the central control unit. Data provided in this way – sometimes dubbed 'proprioceptive' data – again contributes to the intelligent operation of the system. Programs are included to aid the interpretation of the various inputs, and hardware is structured into the robot anatomy to ensure that the necessary data is collected and fed on for processing. Use is made of encoders, potentiometers, tachometers, back-pressure sensors, resolvers and the like – so that the robot brain will always have abundant knowledge of the state of its own anatomy. Biological systems, too, exploit proprioceptive data in many ways, but here – as befits species based on hydrocarbon substances – the sensors invariably have a biochemical component (the chemistry of the biochemical five senses is today well-understood). Robot sensors, by contrast, can be designed and built in many different ways. Typically, proprioceptive data is derived from mechanical and electrical systems, with data from the external sensors derived from 'organs' that include a semiconductor component. Thus 'the vendors are betting their chips on silicon sensors' (Iverson, 1989), a circumstance that signals not only a particular fabrication technology but also the development of systems that adapt well to the evolution of facilities interested in the effective processing of information.

The emergence of robot sensors, systems that often draw on non-robotic technologies, is essential to the evolution of truly humanoid systems. An anatomy that resembles a man or woman, but which lacks intelligence, will always be limited in its anthropomorphic status. Members of the species *Homo sapiens* have minds as well as bodies; the evolution of artificial sensors is helping robots to develop along the same route.

Touching Scenes

It has long been thought useful to provide robots with a sense of touch. For example, tactile sensors were included in many of the early gripper designs to signal the condition of 'over-grip' (when gripper fingers try to grasp a non-existent object). More recently, efforts have been made to design artificial skin, not merely as a protective covering but also as a sensitive surface able to signal awareness of physical contact with items in the outside world. Early Auto-Place robots, common in the 1970s, often included an over-grip detector: when the fingers tried to close beyond a specified 'normal' position, it was assumed that the item was missing – it might have fallen from an assembly line, or the end of a batch might have been reached.

Vacuum effects may also be exploited to provide robots with an artificial tactile awareness. When a robot picks up an object there may be a change in the surrounding air pressure: this may be used to generate a signal to indicate that the action has been completed. The signal may be used in turn to alter the program that is supervising the overall operation. In anthropomorphic terms, the robot notices when an object is absent and then decides to take appropriate action.

Early efforts to design assembly robots often entailed providing systems with a rudimentary sense of touch. This option has been researched in various ways; as, for instance, in the development of robots able to insert a close-fitting peg into a hole, a common requirement in assembly operations. One solution to this kind of problem (considered by Astrop, 1979) is to provide a robot gripper with a range of tactile sensors that can detect the various resistances to the peg entering the hole. When the hand becomes aware of such resistances, signals are generated to make the hand move the peg in an appropriate direction to reduce the resistances to zero. Again efforts are made to duplicate what happens with the human hand: if the robot can 'feel' what is going on, it is in a position to take action to move the peg so that alignment is achieved.

Other research has focused on using 'pulling' rather than 'pushing' the peg to achieve the desired positioning. Early work in the 1970s at the Charles Stark Draper Laboratory in Massachusetts resulted in the development of a compliant device that provided a robot with an effective sense of touch. An arrangement of links was used to define an effective centre of compliance that was coincident with the leading edge of the peg. The peg was still pushed, but the compliant device created an alignment so that an effective 'pull' was achieved (an everyday analogy is the way we manipulate a drawer that has become wedged in its runners). It was claimed that the remote centre compliance (RCC) accomplished 'passively' what more complicated systems achieve with an array of sensors and servo mechanisms.

A distinction is sometimes made between touch and tactile sensing: whereas touch refers to simple contact for force-sensing at only one or a very few points, tactile sensing requires the continuous measurement of forces in an array. The Auto-Place over-grip facility is really the converse of touch sensing, involving the awareness of a missing object. A touch sensor, by contrast, may signal when the gripper is in a position to commence the grasping motion: a gripper may descend on an object, the gripper fingers widely spread, until a touch sensor comes into contact with the object and signals proximity in order that the fingers can be activated to grip the item. Other touch sensors may be deployed to detect burrs on a machined surface, or to precede a welding gun along a seam to be joined. Touch sensors of this sort may be nothing more than simple microswitches or thin probes able to signal contact with a surface or a protuberance. Where spring-loaded probes are activated to move around an object, the changing position of the probes may be used to signal the shape of the object.

Tactile sensing, often a more ambitious affair altogether, involves the design of artificial skin, a sensitive array that can extend over the fingers of a robot gripper, over the articulated arm, or over any other area of robot anatomy that might come into physical contact with the outside world. One of the earliest designs for an artificial-skin sensor was described by M. Briot (1979) at the

Ninth International Symposium on Industrial Robots. Work at the Laboratoire d'Automatique et d'Analyse des Systèmes at Toulouse in France (in parallel with similar research at the Mihailo Pupin Institute at Belgrade in Yugoslavia) had yielded a number of practical devices with possible applications in medicine, industry and other fields. The system described by Briot can identify the position of a mechanical part with multiple planar equilibrium faces, a characteristic activity of sophisticated robots with a tactile-sensing capability. In the 1970s this skin sensor was built into the fingers of a gripper to aid the recognition of an item during the grasping process. Where a robot is equipped with a tactile sensor combined with decision and control facilities the device can recognize something by touching it – much as a blind human being might – and then carry out an action after deciding what to do.

The Briot sensor consists of a number of sensitive points ('nerve endings') carried on a printed circuit board: the points are square, uniformly distributed and organized in a matrix. To activate the sensor a small voltage is applied across a guard ring around the points, whereupon the electrical characteristics of a coating vary according to the amount of pressure exerted. At every affected test point a current variation is caused, which is then translated into a voltage change that can be interpreted to indicate the nature of the object that the sensor is touching. The imprint left by an object is expressed by analogue information representing a pattern to be processed by a computer. In the Briot system use was made of a Mitra 15 computer to carry out the necessary processing after the required information had been collected. Clot and Stojiljkovic (1977) have described a similar early artificial-skin transducer developed in Belgrade. Such devices – and their more sophisticated descendants in the 1980s and 1990s – typically contain arrays of force-sensing elements ('tactels') set on the fingers of a gripper. In principle a tactile sensor could be spread over the entire body of a robot, providing a true sensitive skin; but such an ambitious approach has not been deemed worthwhile in current designs.

The researcher L. D. Harmon (1982) carried out a survey of available tactile sensors. It was suggested that the sensor surface should be flexible and durable, as befits a proper 'skin'; that as many as two hundred tactels could be sited on a small surface, with individual tactels only one millimetre apart; and that the tactel elements should have a rapid response time. It is obvious that the various designs for tactile sensors satisfy such requirements to different degrees. Some tactile sensors rely on the change of resistance in carbon fibre; others on the resistance changes at the interfaces between silicone rubber cords set at right angles to each other; and yet others employ specially developed compressed materials made out of fine yarn impregnated with suitable conductive substances. In most such designs, the application of pressure results in a change of resistance which can be exploited to produce other electrical effects which can then be interpreted to provide information about the object being touched. And it is also possible to design tactile sensors that rely upon magnetic effects.

One design (Chechinsky and Agrawal, 1983) uses a magnetostrictive nickel-based metal. Here, when a force is applied, the induction characteristics of the structure are altered, which in turn generate voltage changes that can be interpreted. A design described by the researcher J. M. Vranish (1984) uses magnetoresistive materials (exploiting a different sort of effect) such as Permalloy, which experience resistance changes when subjected to varying magnetic fields. In one exploitation of these effects, a practical skin sensor comprises a thin rubber sheet in contact with an array of flat wires etched on a sepcial type of plastic. Pressure on the rubber causes the magnetic field to change, yielding further electrical effects that can be interpreted by computer. Some tactile sensors use optical effects. One, for example, carries a mechanical deflector and an electro-optical transducer: when a pin that is integral with the flexible surface material is moved, there is a change to the amount of light passing between a phototransistor and a light-emitting diode. The resultant current that is generated indicates the magnitude of the applied force. In such a way, information can be collected to define the shape and other features of an object in physical contact with the fingers of a robot gripper.

Artificial-skin sensors have been incorporated in underwater robots and in many industrial systems. When work is being carried out in an unfamiliar environment there are obvious advantages in being able to organize interpretative computation on an incremental basis: computation proceeds using initial data, while further data is being accumulated. It is easy to see how this happens in everyday life. Imagine going into a dark, unfamiliar room and searching for the light switch. You may move your hand along the wall, remembering what areas you have covered, until you feel the switch: then you shift your fingers in one direction or another until you are able to operate the switch in the required direction. During this whole process your brain is processing incoming data until the goal, also constructed by your brain, is accomplished.

Progressive computation of this sort can also be used by robots required to function in environments about which, at the onset of the task, only partial information is available.

At the same time it is helpful to use a technology that is best suited to the envisaged application. A growing range of tactile sensors (and touch sensors) is being offered on a commercial basis, and robot engineers need to be well aware of the available options. Some efforts have been made to compare the various designs that can be used to provide robots with sensory faculties. K. E. Pennywitt (1986), for instance, has compared tactile sensors using different technological options (optical, silicone rubber, conductive elastomer, silicon strain gauge and piezoelectric); and I. Plander (1987) has suggested that the piezoelectric polymer polyvinylidene fluoride (PVF–2) is the best material for the construction of tactile sensors. More recently Joseph Alvite, of the Mecanotron Corporation in Minnesota, USA, has filed a patent application for a multilayered skin to fit robot arms. Here use is made of electrically insulating polyester, coated on one side with a metallic conductor. When an object is touched the outer layer of the skin is deformed, exerting pressure on the layers underneath and causing a small electric current to flow: the greater the pressure, the greater the current. Using such a skin, the force exerted by a robot gripper can be accurately controlled, offering suitable protection when fragile items are being handled. And it is also of interest that the current flow in the artificial skin varies with the temperature of the grasped object – so that the robot could be programmed to know when it was in contact with a human being. This is a facility that has many potential applications. A robot sensitive to human contact would have a number of uses as an intimate companion (Chapter 6); and in the industrial environment such a robot could have enhanced safety features – it would quickly stop operations if it inadvertently touched a human being within its sphere of activity.

Many technologies have been explored as relevant to the design of tactile sensors. In a list compiled for TRC by the consultant Stefan Begej eighteen separate technological approaches are indicated (some of which have already been mentioned):

capacitive	beam interruption
contact conductance	beam path modulation
piezoelectric	birefringence
piezoresistive	deformed mirror

potentiometric	use of total internal reflection
Hall effect	optical fibres
magnetoinductance	pneumatic
magnetoresistance	thermal
magnetostriction	ultrasonic

The list is reproduced here to illustrate a central theme of this book: biological systems have evolved under the sole dynamic of trial and error, whereas robot developments can be shaped by (human and machine) intelligence to exploit a wide range of design possibilities.

We may also expect the design of artificial tactile sensors to benefit from further insights into biological systems. One interesting example is the exploration by Kenneth J. Kokjer (1987) into the information capacity of the human fingertip. Such work has relevance to the design of guidance canes for the blind, medical prostheses, and controls in high-performance machines such as aircraft. The research also reminds us just how robot tactile sensors have to evolve before they can begin to compete with human sensitivities. In another approach (Ghani and Rzepczynsky, 1986), an effort is made to combine tactile and visual sensors to provide a robot with high-level faculties.

The authors, working with a number of collaborators at the University of Newcastle upon Tyne, England, set out to develop a tactile sensor and to integrate it with other sensor systems. This again is a nice parallel to what happens with biological configurations: we all habitually combine tactile and visual data to make sense of our environment. The research focused on developing a robot test vehicle able to pick up randomly presented workpieces and placing them with precise control. The system included tactile array sensors mounted on the gripper, a two-dimensional vision system, force-torque sensors and proximity sensors. The aim was to contribute to the requirements of Flexible Manufacturing Systems that involve the automated manipulation of a wide variety of workpieces.

Each of the tactile sensors had thirty-two columns configured orthogonally to thirty-two rows: thus each sensor could be accessed by one row and one column. The columns are provided as copper tracks on a printed circuit board that serves as the sensor base; and each row comprises a piezoresistive rubber tube carrying a flexible conductor. At any point of compression, changes in resistance occur which can be used to generate further electrical signals to allow interpretation of the nature of the touched object.

The emergence of a rudimentary sense of touch arrived early in the evolution of robot systems. Simple microswitches and strain gauges were quite sufficient to provide a degree of touch input to a robot from its immediate environment. The more sophisticated tactile sensors, often requiring high levels of computational processing, came later in the chronology – as did vision sensors. The capacity for vision is a highly complex biological faculty, well beyond the reach of many naturally occurring species (especially those of the plant world). Considering these complexities, which involve data collection, scene analysis, the construction of three-dimensional images and so on, it is remarkable that robots have any talent at all for visual activity. But the 'seeing' robot is already with us, able to watch human beings and other aspects of its environment. There are, moreover, many pressures working to enhance the visual acuity of intelligent machines.

The Seeing Machine

The visual faculty represents an impressive feature in either a biological or a machine context. Vision involves the rapid collection of large amounts of data that are speedily presented for processing by a natural or artificial brain. And there is the further advantage that information about things at a distance is delivered – so many of the hazards of close proximity can be avoided. Moreover the copious data delivered by the vision system can be used to supplement data derived in other ways. There are obvious advantages in providing many types of computer-based systems with a capacity to see what is going on in the world.

Much research into vision has not been specifically focused on robotic systems. Artificial vision has long been a province of AI research, without the assumption that any practical systems would one day find their way into robots. There are in fact many areas of application where seeing machines could be useful: one obvious one, in these security-conscious days, is to have computer-based facilities able to recognize human workers in a factory or office. For example the Wisard system, developed at Brunel University in Middlesex, England, is able to recognize the human face, a facility that has obvious security applications. Another non-robotic area of vision research – with clear implications for robotics – is biology. Again, insights into animal faculties can feed back into the design of artificial humanoid systems.

'Programs for seeing' exist in many different types of animals: in the various cases – consider the compound eye of an insect, the laterally disposed eyes of a shark, the remarkable visual acuity of a hawk – there are characteristic 'hardware' items and characteristic programs for delivering the necessary interpretations. A 'seen object' – which may be a highly complex moving panorama – has presented emitted or reflected light of a certain frequency which causes photochemical changes to occur in the retina. These changes in turn cause appropriate neural signals to be fed along the optic nerve to the sight centres of the brain. The nerve impulses, variously coded and modified, represent the perceived object: the brain then embarks upon a hierarchy of information processing to construct an image of the object.

Examination of biological eyes has provided clues for the design of seeing artefacts. For example, the retina comprises an array of rods and cones that are equipped to detect a small part of the image presented by the lens of the eye. The subsequent electrical and chemical changes facilitate a range of digital and analogue processes that can easily be simulated by electronics engineers. Nerve cells associated with the retina generate on/off responses and also produce the graded changes that are able to influence the subsequent processing activities. Computational algorithms are performed by the vision hardware to achieve the construction of appropriate visual images. It is clear that the combination of algorithms and hardware is essential to the design of artificial sight systems to be used in robots.

The researcher T. Poggio (1984) has described a sequence of algorithms that first extract edge and contour information from visual images and then calculate the depth of objects in the three-dimensional world. We should remember that data about the world is presented not as an accurate three-dimensional array, but as parallel streams of pulses – from which three-dimensional images somehow have to be constructed. The programs provided in artificial sight systems are able to derive shape from shading, from motion, from edge characteristics, from occlusions and so forth; and from the information supplied when, like many biological systems, the artificial 'eye' is provided with a stereopsis capacity. Workers in AI have long been interested in developing the algorithms that can be applied to simulate what goes on in the sight centres of the higher mammals. One approach in constructing a realistic image of the three-dimensional world involves selecting a space location from the retinal image of one eye, identifying the same location in the other retinal image, measuring their positions, and using the disparity to compute the distance of the selected space location. This sort of procedure can be repeated many times to construct a full three-dimensional image that corresponds with an aspect of the real world. There are many clues in this as to how an artificial visual faculty could be designed and constructed for use in robots and other systems. In the mid 1970s researchers at the Massachusetts Institute of Technology (MIT) were able to develop the stereopsis algorithms needed for binocular vision, and which could be processed by a computer. In fact robots have enjoyed a rudimentary vision faculty for over two decades.

For a robot to be able to see, it must be able to perform many different tasks. These, mirroring what happens in biological systems, include:

- Being able to discern points and areas of different intensity
- Knowing how to group discerned features
- Knowing what features to ignore as superfluous
- Being able to infer details about hidden parts
- Being able to resolve seeming inconsistencies in presented data
- Knowing how to decide between different scene interpretations
- Recognizing that an obvious interpretation may be wrong

Different types of calculations are required for all these tasks, and the list is far from exhaustive. Expectation is a key factor in how human beings interpret visual data: if we see a centaur prancing down the road we are likely to blink and to look again more closely. So it may be important to build an 'expectation capacity' into robot vision – so that the system will quickly hunt for more data if a visual construct is very much at odds with the world the robot knows.

There are many ways of classifying work in artificial vision. Cohen and Feigenbaum (1982), working in mainstream AI, have focused on three overlapping fields of vision research: signal processing, pattern recognition and understanding. This approach represents an effective processing hierarchy: the level of processing accords with the complexity of the results that are expected. Thus at an initial level it may be enough for an input signal simply to be converted to a more useful form; at a higher level it may be necessary to perform complex calculations, as may be required in such activities as scene construction or the building of a moving three-dimensional image. Early vision research grew out of work in the 1950s and 1960s that was concerned with the straightforward recognition of characters for text processing and other purposes. Later work focused on enabling a computer to recognize the contents of a world inhabited by boxes, pyramids and other well-defined shapes. Adolfo Guzman's SEE program was designed to analyse straight-line drawings, focusing on the geometry of line junctions to work out what the drawings contained. Variants of the SEE program, and of the daughter program BACKGROUND, have been incorporated in robots to provide a simple visual faculty.

Other researchers have concentrated on specific aspects of a visual field. Thus G. Falk developed the INTERPRET program to interpret a photograph; and Y. Shirai worked on program systems that could generate lines from intensity arrays in photographs. This latter system is interesting in that an 'expectation capacity' is written into the program: if the system suspects that a particular object is a three-dimensional block it may be encouraged to search for further evidence in the picture that will support the hypothesis; and if the evidence is not forthcoming it may abandon the hypothesis and start again. One early system, POPEYE, described by Sloman (1978), is able to sample the parts of an image until fragments are found that suggest the presence of lines: the program can build on early decisions to construct and redefine an interpretation. Cohen and Feigenbaum (1982) survey a wide range of early vision systems, many of which have suggested how visual faculties may be built into robotic systems.

Any vision system intended for use in an operational robot, whatever the environment, has to be able to overcome a number of basic difficulties (some of which have already been mentioned). How, for example, can a robot see that two parts are overlapping, in circumstances where the constrast is poor and the collected data is partial? The overlapping items may appear as a single object: a human being may not be fooled, but the robot computer may jump to inaccurate conclusions. This suggests that the robot environment should be well illuminated, to make any objects as distinct as possible: it is easy for sighted robots to misinterpret overlapping parts in dim lighting, as indeed human beings too may get confused. But there are also problems with strong lighting: artificial contour lines may be generated by shadows, suggesting to the robot the existence of objects that are not really there.

A number of experimental systems, combining robot actuators and vision facilities, were developed from the early 1970s. One system, designed at the University of Rhode Island, USA, aimed to develop 'general methods for robots with vision to acquire, orient, and transport workpieces . . . to assist in increasing the range of industrial applications' for such robots. Here a robot was developed that was able to use vision to locate and grasp randomly placed items in a bin, to determine the orientation of the workpiece in the robot hand, to manipulate the item, to convey it to a site, and to insert the workpiece into a specified location. Two cameras were provided to offer the vision facility: one was mounted on the robot arm and faced downwards; the other was sited at the workstation. The two devices were aligned with the robot axes of motion to simplify the conversion from the coordinates of the cameras to those of the robot. And the system had a degree of intelligence: if the robot positioned a workpiece so that the camera could not obtain a good view, the robot shifted its arm and waited for the necessary instructions. This system, developed during the mid 1970s, was one of the first that were able to feed randomly oriented items from a container to a defined location.

Another early system was based on the CONSIGHT lighting arrangement, in which a narrow line of light was projected across the surface of a conveyor belt. In a typical use of the facility, a robot was controlled by a perceptual system able to recognize parts passing along the conveyor. At the end of an operation the robot asked for further instructions, duly supplied by a computer able to accept inputs from the visual system. Another system used a strobe light shone across an object: the robot was able to detect the object as a narrow line of light. The computer could tell whether the robot was in a position to grasp the object; if not, the computer shifted the robot to a new position and invited it to try again!

In the early 1970s Japan was taking out patents for tactile and visual sensors, and many of these devices were duly incorporated in the robots that were designed and built in the years that followed. Various types of vidicon cameras were used by Hitachi to recognize shapes in die bonding and to inspect nuclear power plants; and by such companies as Mitsubishi and Kawasaki for inclusion in industrial robots.

At the same time research into charge-coupled devices (CCDs) was yielding efficient robot 'eyes'. Microelectronic systems, including CCD silicon chips, were designed at Hughes Aircraft in the

US to carry information on 'packets' of electrical charge, a procedure that could be adapted in the design of robot eyes. Such devices could construct an image in about fifty milliseconds, much faster than a digital computer could manage. It was shown that robots using CCD eyes could interpret the outlines of objects almost instantaneously (*New Scientist*, 29 November 1979). The CCD camera is essentially a miniature array of light-sensitive cells carried on a silicon chip. Such a device can convert an image into a binary pattern in a single stage, and then feed it to a central processing unit (CPU) for interpretation. As with biological systems, it is not the eye that sees, but the brain.

Some early robot systems were built to follow a light source and the movement of a robot arm – so that control data could be fed to the central computer. It was soon found that the processing of visual data took time, which affected the speed with which the robot could respond in changing circumstances. One approach was to limit the amount of visual data being fed to the control system: this helps to reduce the amount of memory needed, and to speed up the processing activity (for example, a computer is wasting time if it insists on processing a mass of data about the background to the task in hand). Here the aim is to activate enough processing for the required operation, but no more. A robot may, for instance, only be interested in whether an object is present or absent: there is no point, in such circumstances, in working out details such as the shape of the object, its precise orientation and whether it carries burrs. Where detailed information about the object *is* required, then additional processing is unavoidable.

In broad terms there is a four-level hierarchy of discrimination, each level requiring more complicated computer processing:

- Detection (is the object present?)
- Position (how is the object orientated?)
- Identification (what is the type of object?)
- Understanding (what are the implications?)

The amount of processing that is feasible depends upon the quality of the camera equipment (how much data can it collect?), on the speed of operation of the control computer (how much processing is it able to perform?) and on the amount of circumstantial detail that is available as background information (how extensive is the computer database?). There are many processing options: the amount of processing that is designed into a system should relate specifically to the task to be performed (in today's world, a matter of simple economic expediency).

Many early vision sensors relied upon relatively simple black-and-white cameras (television cameras were often used) – to provide a robot computer with an array image of brightness levels, the resolution of the camera determining the accuracy of the two-dimensional image. The task was to find objects in the array, and to distinguish them from the background and from each other. Use was made of 'edge finding' (where points found on contrast edges were linked together) and 'region growing' (where it is assumed that the image of a surface is uniform in its local properties). Such techniques represented a first stage in the determination of the properties of objects – which in turn enabled the robot to be instructed to behave appropriately.

At the 1977 Autofact ('automated factory') exhibition in Detroit a number of companies presented robots that could see. An Auto-Place system was designed to load a cup with dice, toss them out, and then search the surface for the dice so that the procedure could be repeated. This robot, the AP-C2, was a clear precursor to many of today's robots that have a vision capability. The system relied upon General Electric TN2000 solid-state video cameras and on microcomputers (the Imsai 8080 in the prototype version, and the Intel System 8020 for production models). Auto-Place robots were equipped with the Opto-Sense vision facility in 1976 as a standard feature, and the SIGHT–1 system – used to find non-overlapping parts – was introduced at the General Motors Delco Electronics Division in 1978; this latter facility was regarded as one of the first 'seeing' systems operating on an automotive production line.

One of the first industrial robots to use a television camera was the SIRCH assembly robot at Nottingham in England (described by Heginbotham et al, 1973). Here a four-position turret carried a lens and three different grippers, and could be moved in three dimensions over a table. The computer – a primitive device from prehistory, by modern standards – was a 12K core store system, able to interpret information collected from the camera field. Once an object had been identified in a broad scene, the manipulator head of the machine was lowered over the object to enable the resulting image to be compared with a reference area in the memory of the computer. No attempt was made to infer the three-dimensional properties of the object being scrutinized.

Later systems have benefited from both the rapid advances in a range of relevant technologies, and further insights into how vision systems operate in biological species. For example, work has been done on the neural processing of visual data in the mammalian cortex, with continual emphasis on the central task of information processing. The researcher P. Bartlam (1981) begins a discussion of electronic sight with the words: 'The eye is an information-gathering mechanism' – a seemingly obvious comment in the culture of today, but a point that had to be emphasized in the early days of vision research. The combination of biological insight and technological innovation has yielded a wide range of practical systems. For example an issue of the journal *Assembly Automation* (February 1982) carries descriptions of a vision system used to assist in the automatic sorting of silicon chips, and of a vision system designed to aid an assembly robot. By the early 1980s it was becoming obvious that vision facilities would increasingly be regarded as standard in a growing range of robots – in particular, those devoted to important experimental work, and dedicated devices required to function with a degree of intelligence on assembly lines. By the mid 1980s many practical robots, with standard vision capabilities, were engaged throughout the world in useful industrial operations. For example, vision systems were being used to examine food products, and many other modern robot applications (see Chapter 5) rely upon a vision capability.

Today the central tasks to be carried out in robot vision are well understood. It is necessary, for example, to accomplish such operations as image acquisition, image processing, object recognition, depth analysis and true visual comprehension; and these various tasks require modifications and enhancements, according to the defined objectives. In one approach highlighted by Joseph Engelberger (1989), a particular range of activities is essential if a vision system is to accomplish useful understanding. It is necessary, for instance, to carry out image preprocessing for enhancement purposes (this involves instructing the sensors, processing pixels (picture elements) and storing data); image extraction to aid object recognition (involving necessary computations and the generation of appropriate instructions); image analysis (defining data structures and object relations, acquiring knowledge and generating models); and image context analysis (exploiting stored knowledge, reasoning and asking questions). Again a hierarchy of activities is encountered, a nice depiction of what various species of robots have been able to accomplish at different stages of evolution.

Researchers have traditionally given attention to such vision topics as grey-scale, stereo and structured light procedures, adapting what is known about the physics, chemistry and biology of vision systems to aid the design of sighted robots. Grey-scale systems – requiring sensitivity to variations of light intensity – have been incorporated in many practical robot systems; for example, in the PUMA robot used in the University of Rhode Island robotics laboratory to pick connecting-rod castings from a bin. However, grey-scale systems, requiring the allocation of many different pixel values, are often too slow for modern industrial processes.

Stereo systems, akin to what is widely encountered in the biological world, offer many advantages but require complicated programming: a computer is required to interpret data from two cameras and to coordinate the two complex interpretations. The computations are necessary if sophisticated depth analysis, the construction of a true three-dimensional understanding, is to be accomplished. The structured light approach – as, for example, with triangulation – can produce detailed information about an object. Here a light source (typically a laser) is used to provide geometric details about the object: the camera accepts reflections from the illuminated object and provides data upon which the robot brain can work. In such a fashion the robot can build up knowledge about the object's position, orientation and surface features. The use of structured light is problematic if there are other light sources in the vicinity, or if the triangulation cannot be controlled with sufficient precision. And there is also the point – of interest to people wanting robots to be made 'in man's image' – that the structured light approach is far removed from what happens in the biological world. It may be preferable to focus on the familiar stereo vision systems with which most of us are well equipped.

Today much research is being carried out into artificial vision, not all of it with robot applications in mind. A collection of research abstracts is offered in, for example, the monthly *Abstracts in Artificial Intelligence* published by the Turing Institute in Glasgow, Scotland. Recent items, published in the early 1990s, deal with such topics as shape decomposition, the use of range imagery, the requirements of three-dimensional vision, aspects of feature extraction and the development of stereovision for robots. As with other areas of research, many different aspects are being brought together to encourage the evolution of sophisticated robot faculties.

Artificial Ears

It has been seen that robots are evolving a sense of touch and a range of visual faculties; it is not surprising that they are also developing artificial 'ears'. If robots can hear, and if they also have a computer brain, they are able to make sense of sounds in their environment. Some sounds may signal an unexpected danger; other may be aural instructions delivered by a human being. The provision of artificial robot 'ears' makes it easier for robots and human beings to work (and play) together.

A robot can be designed to receive an aural input via a simple microphone: the problem then is to make sense of what is being 'heard'. At the simplest level a robot could be made to respond to a loud noise in the environment – say, a sudden crash or a loud siren. This would only be a matter of organizing the robot to act if it encountered sound above a certain decibel level. But there is little scope here for intelligent behaviour: human speech, for example, involves much more than simple variations in sound volume.

Most current robots with an aural capability can be trained to recognize the voice of a particular speaker, or of a small number of speakers. The random babble of a crowd would be totally unintelligible, as it would be to most of us. It is also difficult for a computer-based aural unit to make sense of continuous spoken speech: the robot has a happier time if the words are separated by a small pause. In addition, most functional systems can only recognize at best a few hundred words – and not always with 100 per cent accuracy. Speech recognition systems have a long way to go.

One approach is to convert the microphone output into a binary pattern that can be interpreted by the computer. This is accomplished by comparing the binary pattern with patterns already held in the computer memory; where a match is found, the robot can be instructed to act accordingly. An alternative approach, designed to speed up the processing task, is to extract key features from each spoken word by means of dedicated electronic hardware. Here attention can be given to such elements as vowels, fricatives and silences. The results for individual words are more succinct than when a full pattern-matching procedure is attempted; and the subsequent comparisons with 'dictionary' entries can be carried out more quickly.

The principles of aural pattern recognition are much the same as those for the recognition of visual arrays. In fact a spoken word can be displayed on a screen as a two-dimensional visual image (as a 'sonogram', a wave pattern showing sound variation over the duration of an utterance). An electronic 'template' can then be placed over the electronically stored pattern to ascertain the degree of similarity: close similarity indicates that a word has been identified. There will never be complete identity between the spoken word and the template, because even the same speaker never utters a word in exactly the same way each time.

By 1980 voice-recognition systems were able to handle only relatively few words; for instance, the Voice Data-Input Terminal from Nippon Electric could cope with little more than one hundred words, spoken in spaced groups of up to five words – and this system was regarded as quite impressive for the day. Today there are various systems that can handle several hundred words with reasonable accuracy. Much research has been directed at the specific programming requirements of voice-recognition systems (for example, Iverson et al, 1982); and efforts have also been made to combine the specific speech-recognition function with other data-processing operations in the system – an obvious requirement if robots are to develop a multifaceted integrated intelligence. A system developed at the IBM Thomas J. Watson Research Center in New York is capable of recognizing several hundred words with an accuracy of around 95 per cent. The system divides each individual word into sixteen time slices and performs thirty-two computations on each of these. In this way a pattern is generated that can serve as a reference template stored in memory. And researchers at Carnegie-Mellon University in Pittsburgh have developed the 'Harpy' system, able to cope with complete sentences drawn from a total vocabulary of one thousand words. In one test, Harpy was able to recognize sentences spoken by five different people in a typical room.

More recently an intelligent voice-activated robot arm has been described by Dinash Mital and Goh Wee Leng (1989), working at the Nanyang Technological Institute in Singapore. It is recognized that speech recognition is emerging as a key 'man-machine interface medium': how easy it would be if we could just tell robots what to do, as we might an uncomplaining slave, without having to bother with teach pendants, complicated programming and all the rest! Robots would then be able to interface easily with human beings, involving themselves with social intercourse at every level.

The system described enables the robot to interact with a human user by employing terms that are familiar to most people. Because of the limitations of current technology there is particular focus on the use of short spoken-English sentences: with such a restriction, 'connected speech recognition can be implemented with a good success rate'. The robot indicates that it has understood the spoken words by acting in the appropriate way: in particular, the robot is designed to handle blocks, picking them up and moving them from one place to another without collision. Use is made of an IBM PC/AT microcomputer, the MICROEAR voice-activated hardward and a Sorbot ER-III robot. The control software was adapted to handle speech-activated commands, and to improve the performance of the robot.

There are many software sub-modules, but the following are worth mentioning:

- PUT-ON This sub-module identifies one block as a target and another as a destination. Then one block is placed on top of another
- PICK-UP This moves the robot arm to pick up a block, and transfers control to the next sub-module
- MOV-ARM This moves the robot arm to the position of the destination block
- UN-GRASP This causes the robot to release the grasped block, other sub-modules already having positioned the robot

As many as sixty-four voice commands have been used successfully in the experimental environment. The robot is bright enough to remember the position of all the parts in the working area. The authors, having exploited a range of AI procedures, declare that the system is 'user-friendly and intelligent'. It is emphasized that there is scope for expansion of the sytem's competence: the '64 words in our implementation . . . can easily be expanded to 256 words'. There would be much scope for a mobile robot able to obey more than two hundred spoken commands in the human environment.

Smell and Taste

Little research attention has been given to the development of a robot sense of smell, though its possible use for robot chefs has been discussed! There are, however, a number of circumstances in which it would be useful for robots to be able to detect gases floating in the air. Toxic substances often have a characteristic odour, and the presence of a fire is likely to be signalled by smoke. A robot able to smell smoke and poisonous gases would clearly have a number of useful applications.

It has long been known that gases surrounding a semiconducting material can affect conductivity properties, and this fact can be exploited in the design of smell sensors. In this way robots may come to be equipped with a sense of smell by virtue of semiconductor systems that can register the presence of a wide range of gases. It is said that human beings are able to distinguish around four thousand scents, but there are a few insights into how this is accomplished: the robot sense of smell, based on the sensitivity of semiconductor materials, may be forced to follow its own evolutionary route. A number of devices able to detect gas concentrations in parts per million have already been designed, but such artificial 'noses' are crude compared with their sophisticated mammalian counterparts.

It is clear that fire-fighting robots, and robotic systems intended for use in chemical plants, could usefully be given a sense of smell. In the United States research is being conducted into the development of sentry robots for use in state penitentiaries; here, part of the required design specification is an olfactory sensor capable of detecting the faint trace of ammonia left behind by a human being. As with the other senses, a smell sensor would not be expected to operate in isolation from data derived from other sources. It would be useful for a fire-fighting robot that could smell the smoke to be able also to see the flames: the random human smoker may not welcome the rapid response of a highly automated fire-fighting system!

Many different types of robots are already used in the food industry, mainly for manufacture and packaging – how helpful if they could sample products to determine quality. Similarly, robots with sensitive noses could well be useful in the pharmaceutical industry, and in medical environments where the use of drugs is commonplace. And what is true of smell is true, *mutatis mutandis*, of taste: the food robots could taste the chocolates and fish fingers as well as smelling them! Robot wine tasters might be able to outperform human experts; and the manufacturers of cheeses, cakes, biscuits and new synthetic food concoctions may have to cope with retail outlets staffed with humanoid artefacts equipped with the capacity to smell and taste the products (by that time the manufacturers too will probably be robots!).

This chapter has focused on some of the principal elements that combine to provide modern robots with their capacity for intelligent behaviour in the real world. It has been emphasized that true intelligence is linked to the concept of autonomy: a system can only be self-moving (or 'self-motivating') if it has access to data – about itself and its world – that can be intelligently processed to help the system achieve defined goals.

Intelligent systems characteristically receive their data from sensors of various types ('proprioceptive' sensors provide data about the state of the intelligent system itself). This data – collected by the familiar five mammalian senses, or by senses that only characterize intelligent artefacts – is fed to a central processing unit (CPU) for manipulation by a spectrum of programs that characterize the overall system: a human being is equipped with one stack of programs, an intelligent keyboard-playing robot with another. The more complicated the program spectrum, the more versatile the system behaviour. The simplest robots, perhaps equipped with primitive touch sensors, can only perform routine tasks that admit no unexpected events and no autonomous shifts to new roles; the most sophisticated robots can, like human beings, carry out a range of tasks in parallel, take decisions on the basis on fresh incoming data, and shift speedily to new behavioural trajectories when this is required for the accomplishment of defined objectives.

It has also been emphasized that biological research is often highly relevant to the progressive design of intelligent robots. There are many features in human beings, and in the higher mammals, they may serve as evolutionary goals for intelligent artefacts (see also Chapter 7). But it should not be assumed that robots will not find their own evolutionary routes, unconstrained as they are by the circumstances of biological evolution. The evolution of sensory equipment is a case in point. Biological systems have exploited a few natural phenomena to evolve the five characteristic senses: use is made of parts of the electromagnetic spectrum, of molecules floating in the air, of how contact pressure can translate into coded electrical signals, and of vibrations in the atmosphere. In principle it is possible for robots to exploit any natural effects in the design of artificial sensors. In addition to the effects used by biological systems, robots can 'run to extremes' – using the far reaches of the electromagnetic spectrum, the slightest oscillation of the air, the rarest molecule – to gather sensory data; and effects scarcely noticed by biology – magnetism, electrical static, solar radiation, terrestrial nuclear discharges, vacuum effects and so forth – may come to serve as the bases for a new family of robot sensors.

Once the sensors have delivered the data it needs to be processed – bringing us at once into mainstream computer technology. Today massively increased computer power is being packed into ever smaller volumes, encouraging the progressive evolution of on-board robot brains. Research into biosensors (Lowe, 1985; Roef, 1987), optical sensors (Hartog, 1987) and into many other associated technologies will further expand the scope of intelligent artefacts, reducing the gap between man and machine as robots evolve in the direction of *Homo sapiens*. New advances in computer science – in particular, in such areas as artificial intelligence (AT) and neural networks – will make it increasingly realistic to talk about the mental lives of robots.

Some practical robotic systems have been mentioned in this chapter, with reference to a few application uses. It is helpful now to survey the broad class of robot applications, to see just how far some intelligent machines have already travelled in a few years of evolution.

5 The Applications Spectrum

All computer-based systems, like all machines, are intended to *behave* in various ways. The behaviour may be relatively mundane and uncomplicated: for example, the system behaviour may comprise nothing more than a shifting display of symbols on a screen. Many modern robots are interesting precisely because their behaviour is complex, sometimes signalling intelligence and with obviously anthropomorphic characteristics. Today the applications spectrum for computer-based robots is extremely wide. They are active in factories and offices, in schools and hospitals, at the bottom of the sea and in outer space. We commonly expect robots to be used on car assembly lines, but we are less acquainted with them as waiters, nurses, surgeons and librarians.

There are already many indications that intelligent robots will be able to perform a range of service jobs. Already robot barmen are able to serve up to thirty mixes of drinks in California, robot librarians are working in some Japanese universities, and robot guards are being developed for use in American prisons (perhaps this is what the British Tory Party means by the privatization of the prisons!). In the library of the Japanese Kanazawa Industrial University, nearly three dozen 'intellibots' – small wheeled robots – fetch and replace around two thousand video- and one thousand audiotapes, on demand, for four and a half thousand students. The robots trundle along at a brisk 3 mph or so (5–6 km per hour), select the tapes, insert them into players, and later remove them and refile them correctly in their designated storage locations.

It is highly likely that many traditional service jobs, as well as ones in the industrial sector, will come to be performed by intelligent robot systems. Already there is a Los Angeles drive-in supermarket employing an automatic 'picker' able to receive a computer-generated printout of the customer's order, and then to zoom down the warehouse corridors, under computer control, to fetch the specified items. Some Japanese companies have begun employing robot receptionists, and this chapter will introduce robot butchers, robot sheep shearers and robot window cleaners. It is reasonable to suppose that all human activities are accomplished by performing a series of small, well-defined steps (that is, by performing a program); this being so, it is possible in principle to program a computer-based robot to perform any of the tasks traditionally performed by human beings.

The simplest robots carry out the most straightforward operations in a routine and repetitive way. In human society this has generally meant that industrial activities have always been the easiest to computerize.

In Industry

A Unimate industrial robot, perhaps the first in the world to work on an assembly line, was installed by General Motors in 1959. This event signalled the start of a rapid development that led to a vast population of robots throughout the factories of the world, to talk of the 'unmanned factory', and to consequent anxieties about the displacement of human beings by an automated workforce.

Throughout the 1960s and 1970s conferences and exhibitions were held to display the new-fangled artificial industrial workers. For example, in 1979 the British Robot Association held a four-day exhibition and conference at Nottingham University in England. The widest range of robots ever assembled together in Europe was displayed to thousands of excited observers: more than two dozen automated systems were shown in full working order. The show helped to convince many people that industry was facing a robot future. The displayed systems were designed for many different types of industrial application – welding, spray painting, simple assembly tasks and so on – and they were beginning to use compact computer facilities to provide accurate control and the first rudiments of machine intelligence.

By the late 1970s Unimation and Cincinnati Milicron were US leaders in a field of more than two dozen American manufacturers who had already supplied around 2,500 robots to US industry. The early Unimation models, for example – Unimate, PUMA (Programmable Universal Manipulator for Assembly) and the Apprentice –

were able to perform such tasks as pedestal welding, spot welding, die casting, forging, glass handling, injection moulding, gas welding, machine loading and investment casting. Unimation provided a detailed set of Application Notes, listing dozens of examples of practical applications in operating environments. A typical Unimate robot was programmed using a lead-through technique in which the programmer used dedicated controls to convey the robot through the required operational sequence: once the robot had performed the sequence, it never forgot.

Some early robots, like some modern systems, were based on minicomputers: for example, the CH6 robot from Cincinnati Milicron could be set up to provide a wide variety of tasks under minicomputer control. At the same time there were still many relatively simple robots with little or no (re)programmability: for instance, many industrial operations were being performed by simple Auto-Mate and Auto-Place limited-sequence pick-and-place units. Most robots, as today, were handling relatively straightforward tasks such as paint spraying, welding and sealant application. More ambitiously, 'universal' machines were also being designed and implemented to offer application flexibility and the capacity to perform complicated tasks, such as those involved in the assembly of electrical and mechanical components. Typically, as has already been seen, the early robots were taught what to do: scope for independent robot initial, an element of autonomy, had to wait until the design of artificial intelligence (AI) facilities in the 1980s and 1990s.

By the mid 1970s sensory systems were being incorporated into industrial robots to give them some awareness of their functional environment. Robot miners were implemented at that time, and it was quickly realized that 'seeing' machines would be of particular use in the mining world. For this purpose Professor Thring of Queen Mary College, London, worked on a range of telechiric methods, whereby mining robots could be given sight by means of television cameras in their heads. The aim was to design systems that could be controlled from the surface by human beings wearing helmets in communication with the robot cameras: a human controller would scan an underground scene by turning his own head from side to side to direct the surrogate head on the underground robot, and the robot arms could be controlled in the same way to carry out the necessary mining tasks. Research was also undertaken into systems that could carry out complicated tasks in the factory environment. Soon there were artificial systems able to frame complete car bodies, to assemble complicated electrical equipment, to feed materials and tools to computer-controlled machinery, to handle delicate glass products, to carry out a range of foundry operations and to manage warehouses. Already it was clear that the total manufacturing loop (i.e. all the connected processes involved in making a product) was seeing an invasion of computer-controlled robots able to carry out a wide range of activities – from the initial mining of raw materials, through the various manufacturing processes, to the packaging and warehousing of products at the other end.

In many applications it is useful for the workpiece to be firmly gripped by the robot; in other types of application, the robot holds a tool that operates on a workpiece held in some other way or conveyed on a mobile transport system (for example, a conveyor belt). Examples of this latter type of application include metalworking (for example flame-cutting, grinding, pneumatic chipping), joining (spot welding, arc welding, stud welding), surface treatment (paint spraying, enamel spraying, glass fibre and resin spraying, ceramic ware finishing, applying sealant compounds) and inspection. By the 1990s Unimate options for 'material transfer' – one application sector among many – included, according to press handouts issued by the Unimation Company: '. . . automotove assembly – automotive parts – glass – textile – ordnance – appliance manufacturing – moulded products – heat treating – paper products – plating – conveyor and monorail loading and unloading – palletising and depalletising – an endless list of jobs. . . .'

By the mid 1980s it was already clear that manufacturing companies would have to automate the bulk of their operations or suffer in competition with more imaginative firms (hence the cryptic advice: 'Automate or liquidate!'). Car makers throughout the world were using robots to weld car frames. At the Birmingham, England plant of TI Tubes Ltd, a heavy-duty Unimate robot was installed to handle more than four hundred forgings for automotive axle cases and hub ends in a single shift. Saab-Scania in Sweden introduced robots to resistance-weld parts of vehicles. And in the United States GTE Sylvania used extensive robot technology to automate their transfer moulding machines. Such examples could be almost indefinitely extended: it was becoming increasingly clear that there were few, if any, industrial jobs that could not in principle be performed as well by computer-based robots as by human beings. This realization had many implications for the patterns of industrial investments, the attitudes of trade unionists throughout the world, and the thoughts of system designers working to reduce all manner of activities to the discrete steps that would allow for effective program control.

The idea of the fully automated factory had been introduced as a fictional concept early in the twentieth century. By the 1970s it was obvious that the notion could be fleshed out as a real practical possibility. Already fabrication plants needing minimal human involvement were being designed and built. Neil Ruzic, from the US National Space Institute, described in 1978 the degree of automation at the McDonnell Douglas plant in St Louis ('The Automated Factory – A Dream Coming True?'). In this factory two dozen acres (9.7 ha) of milling machines were used to grind grooves and slots into airframe parts to an accuracy of 0.0025 in (0.064 mm), and 'The machines, for the most part, work alone – watchdogged by only a few men who glance occasionally at a control panel or sweep the cuttings'. The men in the factory are not in charge; that task is left to computers. The automated machines are controlled by numerical systems which in turn are supervised by a hierarchy of computers, the entire hierarchy controlled by a master computer. This extensive plant, covering 750,000 sq ft (nearly 70,000 sq m), was one of the most impressive examples of early automation, but it was not alone. The 1970s saw many automated plant introduced in the USA, Europe and Japan – variously designed to produce cars, engines, earthmovers, building lifts, electrical products and machine tools. Arthur Astrop (1979), considering Japanese developments, discusses how the 'factory of the future is no place for man'. Here it is freely acknowledged that 'The dream of the unmanned factory, where it will be possible to introduce raw material at one end and to pick up the finished, tested and packed engineering products at the other, is becoming a reality.' The means is provided by a range of computer-controlled facilities, including industrial robots; the question was no longer *whether* truly unmanned manufacturing was a possibility but *when* it would be accomplished. A conference in the mid 1980s in Amsterdam, called The Factory of the Future, suggested that it would not be too long.

The Japanese company Toyota was one of the first carmakers to use robots to weld underbodies: a configuration of Unimate 400s were employed to weld sections of the car floor together and a number of cross-members in position. In the early installations about half a dozen Unimates would work in close cooperation, welding electrodes being attached to large jaws so that they could reach to the centre of the panels. Many other car manufacturers were quick to develop robot applications for work on assembly lines. By 1980 the Japanese firms Nissan and Toshiba had developed the Tosman 200 robot, a relatively simple device with three degrees of freedom. Four or five of the robots were able to work together to weld side assemblies and floors; each small robot was only required to make five to ten welds – a far cry from the ambitious machines in modern factories. The Tosman 200 was developed essentially as an economic unit: each device cost about a tenth that of a universal robot.

By the early 1980s a number of British firms were using industrial robots for forging and other applications. Tube Investments, for instance, used Unimates for forging and deep-drawing. At the Birmingham factory of TI Tubes Ltd, an eighteen-billet rotary hearth furnace, designed to work with an Ajax upsetter, was found to be impractical because the capacity of the furnace was too demanding for the three-man team. A robot was the answer – and it was found that a Unimate 4000 robot could easily cope with the furnace and the Ajax. Further trials enabled a Unimate 2005G robot to be introduced in the company's Small Cylinder Division for work handling in connection with hammer necking of high-pressure cylinders produced for fire extinguishers, brewery equipment and medical equipment. In this application the robot was programmed to pick up cylinder blanks from a pallet and to transfer them, in turn, to a furnace, a hammer and a discharge conveyor. Output was dramatically improved and a range of health hazards facing the human operators were totally eliminated. for instance, the persistent back troubles that were common among many furnacemen were much reduced. In the same way the US firm Engineered Sinterings & Plastics Inc (ESP), an early pioneer of hot forging, introduced robots as a matter of economic policy. Because of the conditions inside the furnace area as much as four hours of every shift were required for worker relief: hardly an economic use of expensive plant. The introduction of a programmable robot, able to tolerate the adverse conditions indefinitely, resulted in a massive boost to productivity and the opportunity to take workers out of an unpleasant working environment and to employ them elsewhere. In another forging plant, the fabrication unit of the Alcoa Cleveland Forge designed to produced aluminium forgings, two simple Prab robots and two conveyors were installed. Here one robot was used to transfer the part from the conveyor to the die, while the other was employed to convey the formed part from the die to a second conveyor. The two robots replaced two men in the shift crew, and production rates were significantly improved.

The use of computer-controlled robots in forging and heat-treatment plant quickly became widespread, hazardous and unpleasant environments encouraging the replacement of human beings by machines. Prab industrial robots, for instance, were well suited to use in forge shops, moving blanks in and out of a shot blast machine, thus eliminating the fatigue factor that bore so heavily on human workers. The Prab company developed a special gripper for work of this sort, and other types of end effectors were designed for other dedicated applications. In the late 1970s the Canadian firm International Harvester introduced robots for an aus-forming line. Here aus-tempering takes place – carbon steel is toughened by heating the material to the level at which austenite is formed. In this significant application the robots were able to take the place of three men per shift.

It also soon became clear that robots were well suited to the high-volume runs that typically characterize the production of die castings. By the early 1980s many firms were using robots, in the die-casting environment, to supplement other forms of automation. In the UK such firms as Fry's Diecastings, Metal Castings, Doehler and Wolverhampton Die Castings were quick to exploit the possibilities presented by computer-controlled robot systems. Today robots are used for a wide range of operations in this context: casting extraction, die lubrication, casting inspections, molten metal ladling, insert loading into casting dies, casting quenching, loading castings into (and removing them from) trimming dies, and loading castings on to conveyors and other types of transporters. The programmed sequences are tailored to the needs of the specific operation. In operating a zinc die casting machine, for instance, it is necessary to perform a characteristic range of tasks: the faces of the die must first be lubricated; human intrusion into the area must be prevented; there must be automatic timing of the machine sequence; the masting must be conveyed from the ejectors at exactly the right time; there must be adequate inspection of the casting shot; there must be tight control of the operating temperature, and so on and so forth. In aluminium die casting there is a similar range of necessary tasks. The robot designer will try to develop a flexible system that can accommodate individual user needs: 'universal' designs clearly have the flexibility for many different types of dedicated application, but they are also more expensive than simpler systems that may only be able to perform a limited range of tasks. As with all computer-based systems, there is an inevitable trade-off between machine competence and cost.

Some systems can readily transfer from one type of task to another, allowing flexibility of system exploitation with minimum reprogramming. When the Evinrude Motors Division of the Outboard Marine Corporation began using robots for making ceramic shell moulds – castings used in a range of leisure-type products, from outboard motors to snowmobiles – it was soon found that the Unimation robots could adapt, with little need for fresh programming, from one type of casting to another. Again, the advantages of robots over men were evident: parts too heavy for human handling could easily be manipulated by machines, and robot performance was consistent over lengthy periods where human beings might have been expected to tire and lose motivation.

The 1980s saw a rapid expansion of the number of robot systems able to assemble relatively complex items such as automobile alternators, electric motors and electric typewriter subassemblies. The 'compliance' concept (see Chapter 4) was developed to help the task of bringing manufactured components together for assembly purposes; and a range of sensor facilities was being developed to aid the tasks of the functional robot in the assembly environment. The earliest (pick-and-place) robots did nothing more than pick an item off a conveyor belt and transport it to another location, but later robots became increasingly ambitious, combining a mix of talents to achieve complicated assembly objectives. For example, a robot could be designed to convey electronic components to a printed circuit board, to insert the wires in place and then to make the necessary welded joints: the board could then be automatically conveyed to another workstation for automatic testing, prior to further transportation for automatic packing and despatch.

The Unimation PUMA system, already encountered, was one of the most successful early assembly robots. It was designed to have 'universal' features, making it readily adaptable for the assembly of relatively small products – items smaller than a breadbin with individual parts weighing no more than 5 lb (2.3 kg) each. The overall system, happy to include human operators as system elements, was seen as essentially comprising the PUMA itself, a transfer or conveyor device, feeders for component parts, and human beings. In one application of the PUMA robot General Motors decided to dispense with tactile and visual feedback, but it was soon clear that an element of sensory intelligence was essential to machine adaptability: if a machine cannot perceive what is happening, it is bound to lose a degree of adaptability in the working environment.

Research in the Westinghouse Corporation, funded by the US National Science Foundation, focused on how to convert programmable assembly capabilities into practical industrial applications. Here more than sixty product lines were investigated in terms of their suitability for automatic assembly: it was found that the most suitable products had an assembly time of less than 0.3 hours and a batch size of one thousand or less. It was also helpful if the assembled parts comprised fewer than eighty components. A product line was investigated in connection with the assembly of small motors, product design changes being made to simplify difficult wiring operations – so revealing an important principle in the applications of industrial robots. If a robot cannot manage a task then it may not be necessary to make the robot more intelligent: it may be quite sufficient to redesign the task itself, achieving the same product goal by a simpler route. The Westinghouse research identified eight significant differences in assembly style in a study of 450 different motor types. This suggested that there was a choice of assembly centre concept to allow for a sequence of assembly stations. Complementary work at the Stanford Research Institute in California showed how vision facilities could be added to aid the assembly procedure. And a remote centre compliance developed at the Charles Stark Draper Laboratory was also used in this early pilot system. Some of the conclusions from the Westinghouse research were important for the later design of robot assembly systems:

- It is important to develop low-cost robots that can work faster than human beings
- Annual volume and assembly complexity should be taken into account when considering the feasibility of an automatic assembly line
- It may be economically desirable to assemble more than one product on the same assembly line
- If automatic testing and packaging operations are added to the assembly line, the overall configuration is likely to be more cost-effective
- Products should be designed to accommodate robot capabilities, as well as vice versa
- There should be a mix of facilities, including fixed and

reprogrammable elements (human beings should not always be totally excluded)

- To justify cost-effectiveness factors all relevant considerations – e.g. inventory and quality – should be taken into account

Assembly systems have their own characteristic features, as well as being able to draw on other types of robot operations such as welding, materials handling, and the application of fluid for lubrication, sealant and other purposes.

It has been seen that the remote centre compliance (RCC), employed in 'peg-in-the-hole' types of tasks, is an important assembly concept. The researchers H. Makino and Y. Furuya (1982) have described how the concept has been extended in the Selective Compliance Assembly Robot Arm (SCARA), a design that is particularly suitable for the manipulator of an assembly robot. The SCARA robot, developed by Professor Makino of Yamanashi University with the support of thirteen Japanese companies, originally had four motors: the position of the shoulder and elbow joints, controlled by two motors, defined the coordinates of the wrist in the horizontal plane; while two wrist motors controlled the orientation and height of the end effector. The original machine, with four degrees of freedom, was confined to table-top operations, which none the less cover 80 per cent of assembly work.

An advantage of the SCARA configuration is that compliance can be adjusted: the system boasts a selective compliance, offering flexibility for a wide range of different assembly tasks. The wrist can rotate, allowing the machine to drill holes and to insert screws: one of the first applications of the Picmat SCARA was the insertion and tightening of the screws in a door sash assembly. Various Japanese companies have built variants on the original SCARA concept. The Sankyo Seiki firm, for example, has developed the SKILAM system that comes equipped with two sets of SCARA-type arms. One application is on an assembly line for the building of television sets; other uses include the insertion of delicate car parts and the construction of optical equipment. A Nippon Electric assembly robot is very similar to SCARA; here robots are used for the assembly and inspection of telephone receiver components, the assembly of integrated circuits and the assembly of computer keyboard switches.

Other assembly robots include the Pragma from Digital Electronic Automation (DEA, in Turin, Italy), the Sigma (from Olivetti) and the SERIE 3 (also Olivetti). The DEA Pragma robot, also known as the Allegro because of its speedy operation, has been manufactured by General Electric under a DEA licence. The GE robot can have as many as four arms, each with up to five degrees of freedom. Grippers with a sense of touch are included to determine if an item is in the wrong position, or absent altogether. Pragmas have been employed in a number of European factories for many years, assembling components in refrigerators and motor cylinder heads. The Olivetti SERIE 3 robot can function as an overhead gantry arrangement with several arms. A group of such robots can be sited over a conveyor to define a number of assembly stations: the device to be assembled becomes progressively more complete as it passes through each station. With this approach various assembly applications have been realized: electronic connectors, engine covers, injection valves, and other mechanical and electrical products requiring the insertion of screws, pins, washers and the like. Again, the provision of a sensing facility is essential to reliable and flexible robot operation. The robot needs to be able to detect the gripping of an item, and such problems as the sudden jamming of an air-driven screwdriver or the presence of damaged screw threads. Without a sensing facility, such tasks could not be accomplished without human intervention. It is often easier to weld items together than to fasten them with screws and washers.

The Fiat Group Company COMAU has developed a number of robot applications for assembly purposes (Cigna et al, 1987). For example, the SMART six-axis robot has been configured with two telecameras providing three-dimensional vision to achieve the automatic assembly of vehicle wheels. The SMART system is presented with the wheels, complete with hub caps and mounting bolts, whereupon the robot grasps the wheel and 'looks at' the hub to determine the spatial position of the bolt holes. The robot then manipulates the wheel so that it is adjusted to the hub; after that it carries out the wrenching operation and checks the bolt tightness by means of torque and angle sensors. Other SMART applications assemble hinges on vehicle doors and carry out spot washing of mechanical components. For door assembly, an automated pallet presents hinges and bolts to the robot while a second automated pallet presents up to fifteen doors. A robot 'eye' ascertains the positions of the door holes, and the robot brain works out the position coordinates; the system then aligns the holes and bolts, the bolts are screwed down, and an inspection is carried out bolt by bolt using torque and angle sensors to determine that each hinge is correctly assembled. As many as three different types of door can be handled at the same time.

Welding has long been one of the commonest robot applications in the factory environment. By 1980 the British Oxygen Company (BOC) had created a new operating unit, BOC Automated Welding Products (AWP), intended to market welding equipment that embodied state-of-the-art technology; in particular, for the purposes of this book, BOC was then investigating the welding possibilities offered by industrial robots and other types of automated machinery. One example was a programmable arc-welding robot marketed by AWP and based on a system built by Hall Automation. This BOC-HAL robot included the basic robot structure, an arc-welding power source, a hydraulic power pack, a water-cooling and circulation unit, and a gas regulator. The human operator was also provided with a remote control unit and a control console carrying microprocessor-based intelligence and a number of memory modules. The console, in typical control fashion, was able to remember all the three-dimensional motions of the robot arm and to take the necessary control decisions.

In 1976 Volvo installed a robot spot-welding line, operated by thirty Unimate systems, at its Torslanda factory. At the time this was the largest robot installation in any one factory in Sweden, a country known for its extensive use of robots (Unimate tests were first carried out at Torslanda in 1971). Throughout the 1970s there was a rapid expansion of robot welding research and applications in the USA, Europe and Japan. An extensive range of robot welding equipment was on show at the 1977 Essen welding exhibition in West Germany, and at the same time robot tests were being carried out by the British Welding Institute, by British Leyland (using an ASEA IRb6 robot), by Saab-Scania in Sweden, and by many other institutes and companies. ASEA had nearly two hundred robots in operation by 1980, about a third of which were employed in arc and resistance welding. The German organization Keller und Knappich (KUKA) then had about two dozen KUKA-Nachi (4000 MIC/CO_2) arc-welding robots in use, mainly in West Germany; in that country robots were being introduced to weld earthmoving equipment at Orenstein und Keppell, and to weld brackets to axle beams at BMW Dingolfing. By the late 1970s Unimation was supplying more than sixty Unimate robots a month, half of which were destined for resistance welding operations in a wide variety of applications.

At the 1977 Hanover exhibition in West Germany many robot

welders were displayed, including the new Cincinnati Milacron IAcramatic microprocessor-controlled six-axis, jointed-arm '6CH ARM' robot for arc, resistance and spot welding; and in May 1978 the Unimation Apprentice robot made its first British appearance at the Unimation Telford site. The apprentice system, intended mainly for heavy plate welding, was expected to find applications in such areas as shipbuilding and the manufacture of excavators, mining and agricultural machinery. In Austria ASEA IRV–6/6 kg machines were being used in the welding of bicycle frames, and welding robots were installed by Piaggio, Citroën, Renault, Daimler-Benz and other companies (ASEA robots were installed by Leyland's Cowley plant at Oxford in 1978).

By 1980 the Fiat company had introduced a number of Robogate lines in Italy, at the Rivolta factory in Turin and at Casino, to manufacture the Ritmo car. These lines facilitated the robot manufacture of the main body assemblies ('framing') and side assemblies. Car bodies were taken by Digitron trolleys from Robogate to Robogate, each a four-post structure with a pair of transverse gantries designed to support a robot; in addition, pairs of robots were sited on the floor. Each Robogate was equipped with four Unimate 2000s, two above and two below; use was also made of CIMAU robots. All the facilities were controlled by a PDP 11/70 computer with another system kept as a back-up safeguard. By 1980, with the help of the automated Robogates, nearly four hundred cars were being built every day. The success of the Fiat company with robot automation first came to the public eye via the television advertisements for the Skoda automobile ('hand-built by robots').

Companies in other fields were also developing robot automation for welding purposes. Nippon Kokan was using robots to weld ships, employing systems that allowed wide flexibility in traversing both flat and curved surfaces; and various US firms have been using robot systems in military applications such as the welding of missile fins. In principle, wherever welding had been carried out by human beings it can be carried out by robots equipped with sensory discernment and the power of intelligent discretion. COMAU robots, SMART models, are now used on the Fiat Robogates; as they are in today's Volvo plant in Ghent in Belgium, and for the manufacture of Ford vehicles at the Halewood plant in England. Anna Kochan (1989) described robot investments by Citroën and Peugeot, revealing a range of traditional and innovatory implementations.

In 1989 Citroën invested massively in new automation facilities;

for example, the company spent £276 million on assembly and paint facilities at its Rennes plant in Brittany. The new automated body shop required a workforce of 220 human beings and 109 robots. Assembly accuracy was designed to be less than a millimetre, and assembly lines included 'conformity checkers' whose operations are synchronized with those of the welding robots. As the pressed panels arrive at the assembly lines, the geometric alignment of the panels is checked, and in addition the welding current control is adjusted according to the state of wear of the electrodes. Four main assembly operations – nose, driving position, doors and rear hatch, and engine/transmission and suspension assemblies – are performed off-line, enabling components to be brought together for final assembly.

The final assembly line has four automated sections: at one a team of robots deals with the front and rear doors and the rear hatch; at another, six robots are used to install the roof lining, the dashboard unit and the windscreen; at another there is integration of the vehicle's mechanics, and the installation of the front and rear transmission facilities and the front section incorporating the bumper, headlights and radiator grille; finally, three robots are employed to replace the tailgate on the car, three position the seats, three assemble the wheels, and four others position the front and rear doors. A laser vision system is used to help the robots position the wheels on their hubs; the tailgate positioning robot is aided by a CCD-based vision system; and the final installation of the seats is assisted by ultrasonic sensing.

Peugeot is involved in a £900 million investment in automated facilities at its main production site at Sochaux in France. For example, use is made of 91 spot welding robots from Renault Automation to produce 500 cars in an hour. Before the innovations the plant, using 200 welding robots, was only able to produce 1,230 Peugeot 405s a day. At Sochaux, as at Rennes, there is a 'clean room' painting facility, and a number of fully automated lines for assembly. Use is made of forty GMF robots and automated devices to operate doors and bonnets to enable robots to enter and carry on with their allocated tasks.

Robots are being used for many delicate operations: for example, the assembly of front and rear windscreens, Kuka robots, equipped with laser sensors, are used in various applications where precise positioning and control are required; one Kuka robot is assisted by another automated facility, an auxiliary arm holding a camera and light, in determining the position of holes in the assembly of the steering column. The collected information is fed to the robot brain, whereupon the steering column is manipulated by robot via the driver's door into the appropriate position. Peugeot's aim is to create a site able to produce 1,850 vehicles a day, of both middle- and top-range models, by 1994–5.

For about two decades use has also been made of industrial robots for spray-painting cars and other artefacts. In 1975 the UK Binks-Bullows company was developing the RAMP 4088 computer-controlled robot, which could remember and then reproduce the motions of a skilled human sprayer. If the human operator had made an error in the 'teach-through' sequence, it was possible to program in the appropriate corrections. This robot had the capacity to select one of sixteen different programs to spray-paint products hung at random on a conveyor. A sensing device, fitted on the conveyor just before the spray station, was employed to ascertain which item was arriving, the necessary information was fed to the robot brain, and the robot selected the suitable program.

By 1980 various types of automatic spray systems were in widespread use. Some – fixed gun, continuous coater, reciprocator and so on – had little scope for programmability; and robot systems soon became a favoured option. Only a sophisticated robot system could be taught by a skilled human being, could be assigned sensors to provide information to an intelligent computer control system, and could discriminate coherently between different behavioural trajectories. For example, the microprocessor-based Nordson robot system, based on a robot design from the Basfer Company in Milan, was successfully employed by Fiat to provide a uniform paint finish to consistent quality.

Throughout the 1980s the use of spray-painting robots rapidly expanded, offering highly successful results and taking human beings out of hazardous environments. The DeVilbiss Trallfa robot, for instance, was specifically intended for industrial coating applications. In this case electrohydraulic servo controls were programmed by means of a tape- or disk-controlled electronic system. As usual the robot was taught a sequence by a human operator, and could then repeat every step as many times as was required. During the teaching process the robot's joints were moved, individual resolvers were turned, and phase currents were shifted in comparison to corresponding reference values. In this way it was possible to generate digital data that could be used for control purposes. During the subsequent operation the electrohydraulic servocontrol systems were employed to maintain a close similarity between robot performance and the stored sequency information.

The DeVilbiss device could be controlled in various ways. A single cassette tape reader could be employed when identical parts were to be painted, or when the production requirements involved batches of several different parts: cassettes could easily be changed for different components in batch production. A random program selection facility was provided for the spraying of items that appeared in an unpredictable way: while one part is being painted, the next program is being selected. It was possible to store up to four separate magnetic tape memories, each with a capacity for three or five minutes' spraying time: as many as sixteen different spraying programs could be stored in a single tape bank. Alternatively, control information could be contained on computer-controlled magnetic disks for instant accessibility. Computer robot control (CRC) was quickly seen to offer many advantages such as speed of response, accurate control and flexible operation.

A Trallfa robot installed for paint spraying at Gustavsberg Fabrikker in Sweden in 1969 had worked satisfactorily for twenty-five thousand hours by 1980. The UK firm Rolinx installed a Trallfa robot to apply coatings to batch runs of identical items; while Carron Engineering used a similar system to spray vitreous enamel on sanitary ware, central heating equipment and many other types of domestic appliances. Trallfa systems were used by car manufacturers for underseal application and for the painting of van interiors. Brian Rooks (1989) describes some of the current uses of DeVilbiss surface-coating systems.

Industrial robots are also extensively used in such areas as plastic moulding, machine tending, and the assembly and disassembly of cutting tools. A number of companies have automated their transfer moulding machines in order to upgrade the quality of the manufactured parts, to increase the output, to decrease the cycle times and to reduce labour costs. Series 2100 Unimate robots have been employed with heavy transfer moulding presses, able to operate two presses simultaneously and to perform other functions during the cycle time. By 1980 Prab robots were used to replace operators for part of the production cycle in injection moulding waste bins; and in another application a five-axis Unimate 2100B robot was employed to unload two Farrel injection moulding machines used to make elastomer rubber parts. Here the Unimate gripper enters the press, removes the part from the die using a mix of vacuum and mechanical methods, and then removes the part from the press on to a conveyor.

In all such applications there is the implicit assumption that human beings will be progressively removed from the production loop. This even applies in areas where human skills have been long-lived and traditional. For example, at the Paris Industrial Robot Symposium in 1982 it was suggested (Chirouze, 1982) that shoes could be manufactured by robot; and more recently Torgerson and Paul (1988) have considered how a robot with artificial vision might be used in clothes manufacturing ('Results of these tests show that using machine vision for planning robot motion provides an effective solution for implementing automated robotic fabric manipulation').

It is clear that the old dream of the fully automated factory can now be realized: the central questions to be addressed relate more to investment and the proper safeguarding of human interest than to technological feasibility. In the 1970s there was much talk of the 'unmanned factory'; the researcher M. Eugene Merchant, of Metcut Research Associates Inc. (Cincinnati, Ohio), emphasized (1985) that computer-integrated manufacturing (CIM) – including robot systems – 'is clearly shaping the factory of the future'; and today, in 1992, it is obvious that delays in the full-scale automation of industry have much more to do with social and financial constraints than with any serious gaps in the corpus of technological theory. There can be little doubt that industry – in all its manifestations – will experience increased automation in the years ahead: production loops will be closed by computer-controlled machines, and there will be an inevitable trend towards the marginalization of human beings that has yet to be fully addressed. In 'automating or liquidating', companies will profoundly affect the shape and direction of society in the developed world; the developing world will, as ever, be systematically exploited to sustain affluent countries elsewhere.

In Food and Agriculture

Industrial robots are not only used for welding, assembly, spray painting and so on; they are also involved in handling food, shearing sheep and working in the fields. Robots with a vision faculty have been decorating chocolates for several years (Cronshaw, 1982); and there has long been talk of computer-controlled robots serving as butchers and milkmaids. By the mid 1980s Imperial College in London was experimenting with what was to become the world's first robot butcher. A principal aim of the research, supported by the Science and Engineering Research Council and the Sussex meat-processing firm Glengrove, was to produce a robot able to debone backs of bacon.

Bacon comes in various shapes and sizes, presenting particular problems to any potential robot butchers: the meat varies in shape and consistency, and the bones too are different from one animal to another. Dr Colin Vickery, the project's main investigator, declared: 'We've got three main problems. First, there is sensing where the bones are. Second, there is cutting them out. Finally, there is combining the two in an actual robot.' In fact this oversimplifies the difficulty in designing and building a robot able to cut meat in any useful way. It is clear that any such device would need:

- An array of touch and vision sensors, both to provide initial information prior to cutting and to provide on-going information to facilitate accurate control of the cutting operation
- Intelligence, to enable the appropriate operational decisions to be taken during the process

It is also clear that the machine would have to be able to operate with adequate speed. A robot able to cut meat with complete accuracy, but at a snail's pace, may be a technological triumph but it would be useless as an industrial tool. In the UK alone more than 4 million beef animals are killed every year: robots would have to work quickly to cope with this volume. In 1987 Bristol University in England was given a £180,000 grant from the UK Agriculture and Food Research Council to build a robot butcher, and work is continuing elsewhere. The Institute of Food Research in Bristol is collaborating in the work and is able to provide much information on carcasses: its databank carries details of animal anatomies and the distribution of meat and fat for different species, sexes and ages. A principal Bristol University researcher, Koorosh Khodabandehloo, has been quoted as saying (*Sunday Times*, 29 November 1987): 'As we advance towards more sophisticated robots we'll be involved in more complex information processing and more complex mechanical design for the cutting devices, but the basic components – sensors, visual information and the expert system [see Chapter 6] – will all have to be there from the word go.'

A now celebrated robot sheep shearer was described by the researchers J. P. Trevelyan et al (1982). Research into the automatic shearing of sheep has been supported by the Australian Wool Corporation and carried out at the University of Western Australia in Perth. In this remarkable robot application a large mechanical manipulator holds the sheep on its back, securing it in position and then turning it so that the robotic cutting head can gain access to the part to be sheared. Again, some of the problems encountered by robot butchers have to be solved: all sheep are different, not only in their shape and textures but in how readily they submit to being robot-handled in such a fashion!

The Perth research team has built a detailed 'sheep map' by collecting thousands of anatomical measurements which are then fed into the computer for analysis. With access to this information the robot cutter is fed into position and guided by sensors to shear the sheep at exactly the right angle. As the animal breathes, so altering its shape, the sensors detect the change and instantly alter the position and direction of the cutter. More sensors at the tip of the cutter are designed to ensure that the robot cuts only wool and not sudden lumps of flesh. Of course we may expect that the robot will make mistakes from time to time, to the detriment of the unfortunate sheep – but this also happens when human beings are doing the shearing. Human shearers typically work at a price per head, and they often leave long, bleeding wounds (called 'shoe stringing' in the trade). When this happens a 'tar boy' apprentice brushes hot tar on to the injury, and the shearer quickly moves on to the next victim. The sheep may learn to prefer robots!

Efforts are also being made to produce robot milkmaids. A research team at the Agriculture and Food Research Council Institute of Engineering and Research near Bedford in England has developed the necessary theory and carried out a number of practical trials. One of the principal researchers, Mike Street, has developed a milking parlour with robots in mind; and one idea is that such facilities could be provided in the fields where the cows habitually graze. Cows would go up to milking robots when they required relief, and the robots would recognize the cows by means of identity tags. The robot would then apply a vacuum cluster of four cups on to the teats, sense when the milk was running dry, remove the cups and then clean up. The milk could be collected each day without the need for the cows to be brought back each day for milking.

Other robot applications in the agricultural environment have been described by the researcher Fred E. Sistler (1987) of the Louisiana Agricultural Experiment Station at Louisiana State University in the USA. He notes that intelligent systems are being used in grading and packing fruit, and in many other applications. Today artificial vision systems are extensively employed in food processing; as, for example, when broken yolks have to be detected in commercial egg-breaking machines. Robot 'eyes' are also used to detect faulty potato chips (french fries), to sort and grade cucumbers, to inspect chocolate candy coatings and to detect defective pizza crusts.

The Agricultural Engineering Department of the Louisiana Agricultural Experiment Station has also developed a robotic seedling transplanter. The robot, mounted on a drawbar and pulled by a tractor, has an effective robot arm equipped with a gripper. It picks seedlings from horizontal trays and places them in a commercial transplanter: the robot was found to miss an average of one plant in each 36-plant tray, and it worked much more slowly than a human being. Research in Japan (Kondo and Kawamura, 1985) has focused on robotic systems for harvesting fruit from trees (a video camera guides the robot gripper to the fruit); and on an automatic grain combine (Kawamura, 1983). In this latter system a human driver makes the first pass around the field, so defining the area of operation; then he dismounts and the machine completes the rest of the field on its own, only stopping when its grain tank is filled or the entire field has been harvested. The robot, equipped with tactile sensors, can detect the edge of the last swathe, collecting information that can be exploited by the automatic control system. Other Japanese work has been directed at the design of robot sprayers for applying hazardous chemicals to trees. Here, photosensors may be used to enable the robots to sense the presence of tree trunks; and others may be guided along PVC pipes previously set in the ground.

Robots have also been employed as waiters in fast-food restaurants. Joseph Engelberger (1989), for example, mentions the use of mobile robots in Japanese restaurants ('. . . robots roam among the booths of a restaurant, both taking customer orders and making deliveries'). It is obvious that if Japanese robots can fetch cassettes for students in libraries, they can just as easily collect meals from kitchens and deliver them to hungry customers.

In Healthcare

In the early 1980s there was growing talk about how robots might be used in a range of service applications. Brian L. Davies (1984), for example, considered how robots might be employed to help the severely disabled, at the same time reminding us that 'today's robots are still largely blind, dumb and daft'. An obvious approach to the use of robotic systems for disabled people would be to attach a powered robot arm to a wheelchair, so that an artificial limb – under intelligent computer control – could be used for a range of necessary tasks. In the mid 1970s the Veterans Administration Prosthetic Center in New York developed a telescopic arm for mounting on a powered wheelchair. The device also included a voice command facility: when the patient said 'slow', the vehicle would rapidly decrease its rate of travel, so that as it approached its destination it could be moved slowly and accurately. The attached arm was designed to be able to assist with feeding, and to turn the pages of a book or magazine; though at the same time particular attention was given to safety aspects – it is, after all, a general safety provision that human beings should not be able to move within the operational envelope of a robot. Davies commented: 'True safety can only be achieved if the arm has such a low force capability that, even if it goes wild, it can be pushed back with your chin – but even this is not satisfactory if the gripper holds a fork and is aiming at the patient's eye.'

Research has been directed at developing the Unimate Puma 250 robot as a useful attachment to a 'smart wheelchair'. This work, conducted at Stanford University in the USA, has also involved development of a two-fingered hand with optical transducers. Similar research, in the Spartacus project, was carried out in 1985 by no fewer than thirty-three collaborating French companies. This work built on research into robot arms developed for the nuclear industry. In Sweden the sophisticated Sven hand has been widely used by children, though users have been very carefully selected. And the University of Adelaide in South Australia has produced a robotic arm that can serve as a feeder, a page turner and a drawing aid.

It has become increasingly obvious that there are many medical applications for present and future robotic systems. In 1989 the Fulmer Research organization at Slough in the UK set up 'brainstorming' sessions between doctors and robot engineers, as part of a study supported by the Department of Trade and Industry (DTI).

No fewer than four hundred possible applications were identified, ranging from microsurgery to intelligent prostheses: some of the ideas are now being developed by researchers under the terms of the DTI-sponsored Advanced Robotics initiative. Again the crucial safety considerations are being taken into account. Robots may be expected to work in close contact with ill or disabled people, and a suddenly skittish or psychotic robot might do more harm than good! Patrick A. Finlay (1989), principal engineer at Fulmer, has considered some of the main uses of robotic systems in the medical environment.

It is emphasized that in the UK alone more than 750,000 nursing days are lost annually because of back injuries, mostly caused by lifting ill or disabled patients. A robot able to perform such lifting tasks, at the same time offering a patient the dignity of privacy, would have many advantages. Such systems may be seen as economic adjuncts to normal hospital systems, perhaps in circumstances where further investment in staff training and employment is unrealistic. In fact the World Health Organization (WHO) has recommended that automation in health care should be regarded as a priority for developed countries.

In the area of rehabilitation the Japanese 'Meldog' guide robot for the blind is undergoing pre-production tests. This is a powerful mobile system with a comprehensive computer memory: it carries in its memory a detailed map of the town, – intelligent sensors equipped to recognize street junctions, signposts and other geographical landmarks. The robot monitors its progress and provides the necessary navigational instructions to its blind user. Thus the human being is instructed by the robot to halt or to turn right or left; and the person need not worry about his or her walking speed – it is Meldog that constantly makes the necessary adjustments.

Japanese researchers are also developing the Transfer Supporting System for the Handicapped, an automated facility that can lift patients and convey them from one part of a hospital to another. Here an automated guided vehicle system carries a board that can be moved between the patient and the mattress; once the patient is securely supported by the board it converts, where appropriate, into a hinged seat for the transport phase. Meals, washing and other hospital items can also be transferred using automated robot systems.

It is perhaps the use of robots for surgery that would be viewed by human patients with most trepidation, though there are in fact a number of procedures that might be better performed by robots than by human surgeons. For example, the procedure known as radial keratotomy (an aspect of opthalmic microsurgery) requires a very precise depth of cut in the cornea. A human surgeon can manage a cut with an accuracy of 100 microns, where ideal control would be to 20 microns. Canadian researchers are developing a robot that would be able to achieve the ideal figure. Already a Puma robot, using information derived from a stereotactic scan to plan its drill trajectory, has extracted tumour tissue from a human patient for biopsy purposes. In France a robot is being developed to aid a human surgeon in repairing a spinal defect. And Japanese researchers have demonstrated a robot able to transplant a cornea, though not yet on a live patient. Finlay (1989) also highlights the possible use of robots for prostate surgery, an application described in *The Independent* (London, 30 January 1990).

A team at Imperial College in London has demonstrated a robot carrying out a simulated prostate operation (Finlay: '. . . it is difficult for human surgeons to keep track of where they are in the sequence'). The scientists used a potato for the prostate, inserted a lance down an artificial penis, and investigated the robot's ability to cut the appropriate shape in the tissue of the potato. A human surgeon was able to observe the procedure by means of a video camera. Here it was found that the robot, with a blade speed of 40,000 rpm and an excellent comprehension of three-dimensional tasks, achieved an accurate operation in minimal time: human surgeons usually take about an hour to perform a prostatectomy, whereas the robot took five minutes. Michael Coptcoat, a senior registrar in urology, was quoted by *The Independent*: 'Robots will never do difficult operations as well as experienced surgeons. But they should be able to do *simple* operations better then inexperienced surgeons. A TURP [trans-urethral resection of the prostate] is one of the few operations that could be done this way, as the operation involves very specific, repetitive movements.' Another urologist, Hugh Whitfield of St Bartholomew's Hospital in London, believes that patients will accept operations performed by robot once they are shown to be safe. Coptcoat believes that robot prostatectomies will be performed on living patients before the end of the 1990s. (For a discussion of how developments in robot science may assist impotent men, see Chapter 6.)

Scientists at IBM are collaborating with the University of California School of Medicine to develop robots that could assist surgeons in various ways. For example, work has shown that a robot, using synthetic bone for a simulated hip replacement operation, can achieve a more precise fit and accurate allignment of an implant than can a human surgeon working with the usual hand-held instruments. William L. Bargar, an orthopaedic surgeon at the University of California at Davis, has commented that: 'The robot-assisted surgical technique may significantly improve the results of this surgery . . .'. The robot would be provided with detailed information about pin positions to facilitate accurate robot placement of the implant. Finally the robot, under computer control, would drill the implant cavity. It has been suggested that this type of research will also yield robotic applications for plastic surgery, cancer and head and neck surgery.

In June 1990 a robot surgeon called Robodoc carried out its first operation on a live patient: it was used to grind the leg bone of a dog very precisely in order to assist with a canine hip replacement operation. This task – a development of the effective collaboration between IBM and the University of California – was being carried out for the first time by an artificial system (*New Scientist*, 23 June 1990): robots had assisted in operating theatres before, but had never been allowed to cut into living tissue. Use was made of the Orthodoc program, designed to indicate the best position for an artificial joint: the robotic arm, guided by computer intelligence, cut a hole in the bone using a sharp rotary instrument. One of the researchers, Hap Paul, declared: 'The robot is very steady and we are able to program the robot with the exact dimensions of the prosthesis so that we can get a perfect match to the bone. That's impossible to do manually.'

More than 160,000 hip replacements are performed on humans every year in the United States, with a similar number taking place in Europe (in the USA a thousand hip replacements are performed annually on dogs). In such circumstances there is obvious scope for artificial systems that might be able to improve on human performance; and it is easy to speculate about many other possible applications for robot surgeons (Paul: 'I'd like to see it used in knee replacement, and some neurosurgeons are interested in using it for spinal surgery'). A UK group – comprising Fulmer Research, the SAC Hitec automation company, Imperial College and the London Human Computer Interaction Centre – is currently exploring how computer-controlled robots might be developed to perform brain surgery. The robots would be supervised by computer programs that encapsulated the expertise of real human surgeons: the computer would first plan the operation from scan information, and then the robot would be able to perform close to vital regions of the brain where human surgeons may fear to venture. It is also suggested that the robot surgeon would be able to operate with a delicate sensitivity impossible for human beings. In fact a number of successful brain operations have already been carried out by robot surgeons. Thus Engelberger (1989) states: '. . . brain surgery with a robot playing a key role in a neurosurgeon's team has been proven sound in over twenty successful procedures which have been performed on human patients at Memorial Medical Center, in Long Beach, CA'. Dr Yik San Kwoh, an electrical engineer at the Center, began his development work in 1982: the aim then was to produce a robot arm that would be controlled by a computer with access to scan data. The early experiments were performed not on a potato but on a watermelon. Today it is claimed that the robotic procedures at the operating table reduce the surgical time by as much as 50 per cent, so reducing the degree of patient trauma.

Joseph Engelberger has also described the development of robot nurses and robot pharmacists, obvious areas where precise and reliable performance would bring many benefits. Similarly, K. G. Engelhardt, Director of the Center for Human Service Robotics at Carnegie Mellon University at Pittsburgh, Pennsylvania provides a comprehensive overview of the use of robotics in the 'health and human service' areas (Engelhardt, 1989). Here automated systems are profiled under various heads: patient transport/lift/transport; ambulation; housekeeping; physical therapy; depuddler; surveillance and monitoring; physical assistant; nurse assistant; patient assistant; vital signs monitoring; mental stimulation and cognitive rehabilitation; and other (for instance the robots 'currently working in research and diagnostic laboratories'). In all these areas there is immense scope for robot development and implementation (Engelhardt: 'The major challenge in designing these intelligent devices will derive from how acceptable they are to a wide range of users').

It is encouraging to find that there are many actual and potential uses of robots in the 'health and human service' areas. But it must also be remembered that a much greater volume of research – involving prodigiously more financial and scientific resources – is being devoted to the ways in which automated systems, including robots, can be best designed to achieve the effective slaughter of human beings.

In War

When the Gulf War began in January 1991 the world was soon told that computer-based systems had a large part to play. Peter Pringle, reporting from Washington for *The Independent* (21 January 1991), wrote a piece entitled 'Computer is king in chain of command'. Readers learned that the computers in Riyadh, Saudi Arabia, and at the MacDill air base at Tampa, Florida, were working well ('Mistakes in the massive air strikes have been kept to the minimum'). A considerable amount of information was fed into the Templer (Tactical Expert Mission Planner) computer at MacDill, whereupon the system provided the field commanders with a continually updated battle plan. The target cassette tapes for the Tomahawk cruise missile were also produced by computer. Hence there is a real sense in which the Gulf War was shaped in part by 'robot' planners.

In fact much of the equipment used in the Gulf War was computerized. Aircraft and ships relied extensively on on-board computers; and the Patriot missile, much vaunted but of mixed performance, relied upon computer supervision. The Gulf conflict was to a large extent planned and prosecuted by 'robot' warmakers. Military thinkers have long speculated about the role of computer-based systems on the battlefield.

In the early 1980s the American company Robot Defense Systems conceived the fully automated Prowler (Programmable Robot Observer with Logical Enemy Response), a vehicle the size of a tank and able to perform a wide range of tasks under computer control. The Prowler designers hoped that before the turn of the century the mobile system would be able to carry out a number of tasks, including sentry duty, reconnaissance, clearing minefields, operating in contaminated areas, collecting human casualties, attacking armour and defending from air attack. Early versions of the Prowler were radio-controlled, but the ultimate goal is to produce a 'thinking' robot that can negotiate difficult terrain, observe the enemy and undertake the required offensive actions on its own initiative. The robot Prowler would have a memory stocked with infomation about the terrain, would carry a wide range of sensors, and would even be able to determine whether a human body was alive or dead. It is suggested that the Prowler might be equipped with M60 machine guns, 105 mm cannon and Stinger surface-to-air missiles to attack helicopters and other low-flying aircraft. I leave the reader to speculate on the implications of a malfunction in the robot brain.

Many of the design features of the Prowler are now proven in practice. The system computer is well equipped to detect movement and body heat in a battlefield environment, and the associated techniques of pattern recognition, data collection by sensors and high-level decision-making by computer are well established in industrial applications. The Prowler follows a number of other US Army projects – for example, ROBART–1 – for the development of robot soldiers, schemes that consumed massive financial and other resources but which yielded nothing of any practical use. Alternatively the human brain can be left in control of a robotic suit of armour. The researcher Jeffrey A. Moore of the advanced weapons technology group at the Los Alamos National Laboratory in New Mexico has proposed a robotic exoskeleton that could be controlled by brain signals (*New Scientist*, 25 September 1986). The aim is to enhance the performance of the human soldier.

Another suggestion is that future military robots could be miniaturized. Thus Charles Petty, a former Pentagon official, has suggested in *Signal* – the journal of the US Armed Forces Communications and Electronics Association – that robots the size of specks of dust could act as soldiers and spies in future wars (reported in the London *Daily Telegraph*, 1 March 1990). In this case the systems would be based on nonotechnology, the science of sub-microscopic manufacturing where items are measured in nanometres (1 nanometre = 1 thousandth of a millionth of a metre). Researchers in Cambridge, Massachusetts have already built a complete electric motor less than a tenth of a millimetre in size, an intricate device with a top speed of 600,000 rpm. Such minuscule 'robots' could have many uses. Petty observes that 'by the early 21st century machines smaller than pin-heads could power thousands of little robots, patrolling security areas'. And it is easy to speculate on possible military uses: robots no larger than specks of dust would avoid many of the difficulties faced by combatants on the modern battlefield.

Robots of a more normal size can also be used for aircraft servicing: for such tasks as rapid refuelling, weapon loading and decontamination (Engelberger, 1989). It may be pointed out that some of these tasks could also be formed by robots for commercial aircraft, just as many operations performed by mainstream industrial robots are relevant equally to civil and military machines: a robot that makes aeroplanes has no interest in how they are used.

A robot vehicle, designed to substitute for a soldier's eyes and ears in combat, has been described in *The Industrial Robot* (September 1990). The Tactical Multipurpose Automated Platform (TMAP) comes equipped with video cameras and acoustic monitors that facilitate surveillance, reconnaissance and target acquisition. The intention is that a human controller guides the vehicle into a hazardous environment where it uses its sensory capabilities for the collection of battlefield information. Prototype systems have been produced and tested in 'hostile' environments, the aim being to develop 'tele-operations' rather than truly autonomous robot control. TMAP is not the only Scandia project; others include Thomas (TMSS, Telemanaged Mobile Security Station), designed to perform sentry and security operations using radar and infrared sensors; and Fire Ant, designed to detect and destroy moving armoured vehicles up to about 550 yards away. Again such systems are intended to be controlled by human users – though there is no reason in principle why they could not be given programmed expert systems to provide a degree of local autonomy.

In Prisons and for Security Purposes

It is clear that many military robot systems can be adapted for use in a civil environment, as effective guards in prisons and as security devices for company sites and other locations. Already robot guards are in operation, albeit in a largely prototype capacity, in some parts of the United States. Denning Mobile Robotics of Massachusetts has agreed to supply the Southern Steel Company, the nation's largest manufacturer of detention equipment, with several hundred robots a year for several years. Reports of the Denning systems first began appearing in the mid 1980s (see *Infoworld*, 20 February 1984, and *Computerworld*, 7 October 1985), and a number of detailed accounts appeared in subsequent publications (for instance in Engelberger, 1989). It was first suggested that each of the 4 ft (1.2 m) 200 lb (91 kg) mobile robots would sell for around $30,000, though the figure was soon increased to between $45,000 and $65,000. The aim was to provide systems that could be used to augment human guards: for example, to patrol prison corridors at night. The robots are designed to move at about 5 mph (3 km per hour), to detect human odours and to transmit sound and pictures. It has been suggested that such robots would be able to undertake 'suicide' missions in the case of prison riots, venturing into hazardous situations and transmitting information for as long as they survive.

The Robart project, already encountered as a military system, also has relevance to civil security requirements (Everett and Gilbreath, 1988). There are now three Robart systems (I, II and III), progessively equipped with ever more sophisticated technologies. The researchers H. R. Everett and G. A. Gilbreath, from the US Naval Ocean Systems Center at San Diego in California, provide a detailed description of Robart II, a 'supervised autonomous security robot'. Here the design purpose was to provide a multi-sensor, mobile security robot, 'a robust automated system that exhibits a high probability of detection with the ability to distinguish between actual and nuisance alarms'. The robot is controlled by a desk-top computer, and many sensor facilities are included.

The Robart II system can monitor system temperature, room temperature, relative humidity, barometric pressure, ambient light, noise levels, toxic gas, smoke and fire; the system is also equipped to detect body heat, motion, vibration, and sound (motion alone is detected by a complex of optical, ultrasonic, microwave and video sensors). The computer brain estimates whether a real threat exists by combining the weighted scores of alarm-sensor outputs from a given region; if the composite threat exceeds a previously defined level, a true alarm situation is assumed to exist. The robot body is made out of rugged plastic and glass fibre, and the system is propelled by two motors. A stack of internal sensors is included to monitor the performance of the system itself; 256 internal checkpoints are used to check circuit behaviour, the system configuration, switch options, the state of cable connections and other system features. When the system detects that it has an internal problem it generates a speech output to inform its human controllers (not quite 'I don't feel so good. Where's the doctor?', but words to that effect).

What we see here is a range of robotic products that can be used in civil and military applications. The two *Robocop* feature films, popular in the late 1980s and early 1990s, are not entirely fanciful depictions of the future. Today there are already a number of practical security robots being exhaustively tested in a wide range of real-world situations.

In Hazardous Environments

Prisons and wars necessarily create many hazardous situations, but they are not the only ones: there are also environments such as nuclear power plants, undersea terrain and outer space. In all there is a role for intelligent robots.

High-radiation environments present many hazards to human beings; operatives can wear protective clothing but there are still many potential dangers, particularly when lengthy repairs have to be carried out to a reactor or when an entire plant has to be decommissioned. Already various types of robot systems have been used in nuclear power plants for maintenance and decommissioning purposes, and their use is expected to increase in the years ahead. For example, the Rover I remote reconnaissance vehicle was employed to penetrate the radioactive basement of the damaged Three Mile Island nuclear plant in the USA, and to take pictures and radiation measurements. A sibling vehicle, Rover II, then took material samples from the inside walls of containment buildings, so aiding the creation of a comprehensive database of damage information. Another vehicle, Workhorse, has been developed at Carnegie Mellon University in Pittsburgh to take a range of operational initiatives in a nuclear plant environment. Workhorse comes equipped with on-board saws, scrapers and wrenches, all of which can be selected by a 23 ft (7 m) mobile arm. And in the same vein a group of nine Japanese companies, supported by the Ministry for International Trade and Industry (MITI), is working to develop a family of robots for dismantling obsolete nuclear plants. The German company Kerntechnische Hilfdiest Gesellschaft (KHG) has supplied a mobile robot to aid surveillance and sampling tasks at Chernobyl.

Already the ROSA robotic system from the Nuclear Services Division of the Westinghouse Electric Corporation has worked for many years inside nuclear power plants, examining heat exchanger tubes and then accomplishing the necessary repairs. Such a system can work endlessly in an environment that would be fatal to a human being in a matter of minutes. In the UK the Taylor Hitech company has designed a massive robot arm for the Atomic Energy Authority to assist in the dismantling of the highly contaminated internal structures of the Sellafield advanced gas-cooled reactor. A giant dismantling machine sitting on top of the reactor pile will be used to support the 1100 lb (500 kg) arm (*New Scientist*, 22 October 1988). This robotic system is designed to hold cutters, guide grabwires and pick up small items of waste; the grabwires in turn control other automated equipment that will remove contaminted debris from inside the steel pressure vessel. The robotic arm, made out of steel and aluminium, is jointed to allow movements at the wrist, elbow and shoulder. The aim is to dismantle the entire reactor by 1996, at a total cost of some £50 million.

Automatic undersea vehicles are often linked via a cable to a support vessel on the surface; or autonomous robotic devices may 'swim' freely, their on-board brains facilitating navigation and other forms of decision-making. A submersible used by British Telecom International for inspecting and repairing undersea telephone cables can operate in either mode: the 'Seadog' system actually digs a trench in the sea floor, inserts a cable and then fills in the trench and an on-board computer brain provides the system with a degree of local autonomy. Another well-publicized undersea vehicle, in this case remotely operated, was the 'flying eyeball' used to examine the *Titanic* and able to operate at considerable depths. In a similar vein a French underwater vehicle, the *Epaulard*, can work happily enough 18,000 ft (5500 m) under the sea. All such machines are evolving levels of computer-based intelligence that allows independent decision-making.

As in the deep seas, so in outer space: there is already a burgeoning family of robotic systems that have been variously tested in earth orbit and on the surfaces of Mars and the moon. Robot arms have been manipulated by human passengers on the orbital Shuttle, and remotely controlled robotic vehicles (with some local autonomy) have worked well on the surfaces of celestial bodies. The remote manipulator system (RMS) used on the Shuttle, which cost $25 million to develop and build, was designed to lift satellites and other payloads in and out of the cargo bay and to place them accurately in orbit. The arm can reach out to a distance of around 50 ft (15 m) and, though aided by low orbital gravity, can handle masses up to as much 15 tons (15 tonnes). But this is not a smart system: it relies on the power of human brains inside the Shuttle.

Automated lunar vehicles were first successfully used more than two decades ago: people tend to forget that it was as far back as November 1970 that the Soviet Lunokhod 1 landed on the moon's Mare Imbrium, to begin a remarkable exploration. Over a ten-month period the Lunokhod investigated more than 6 miles (10 km) of the moon's surface, sending more than twenty thousand pictures back to earth. This too was a rather dumb creature, persistent but lacking on-board intelligence. The US Viking 1 had more scope, incorporating as it did two automated chemical laboratories, a weather station, a seismology unit, a photographic processing laboratory and a two-computer robot brain. A telescopic arm carried a hoe for digging trenches, and a scope for collecting soil samples and feeding them to the biological and chemical processors. Such systems provide many clues about how intelligent robot facilities will develop in the years to come.

In Training and Instruction

Instructional robots already have a substantial history: they have long been perceived as providing ways in which schools, colleges, engineers and those pursuing hobby activities could experiment with programming languages and control options. The Microbot company in California was one of the first organizations to demonstrate the use of small instructional robots, though other firms were soon keen to climb on the bandwagon. In the 1980s miniature robots were being marketed by such companies as Rhino Robots of Illinois, Mitsubishi Electric of Tokyo and Hikawa of Aichi. John Hill, a founder of Microbot, worked with three associates to develop robot arms that could be programmed on a personal computer (the Radio Shack TRS–80); at the 1980 West Coast Computer Fair in San Francisco there was massive interest: 'We were six people deep and everyone wanted to program the robots. People were climbing on top of each other' (quoted by Asimov and Frenkel, 1985). The upshot was such instructional robots as MiniMover and Teachmover, systems with a maximum reach of 18 in (97 cm) and a top lifting capacity of around 1½ lb (0.7 kg). Other competing systems were the various products from Rhino Robots: the XR–1, the XR–2 (both instructional systems), the Charger (a small industrial device) and the Scorpion (an instructional robot).

One of the best-known instructional robots is the 'Turtle' floor system, designed specifically to aid children learning how to build up complex ideas from relatively simple patterns. Seymour Papert, a main advocate of the associated programming language LOGO, has described the Turtle as 'a device to think with'. In recent years there have been various Turtle systems; and for a while the BBC 'buggy', a versatile floor robot made out of Fischertechnic parts, was in vogue. The small instructional robots and the various floor devices – variously equipped with sensors and on-board computer brains – may be regarded as an economic means to enable both children and adults to gain an appreciation of the many possibilities for advance that lie in the wider robot world.

In the Home

Any self-respecting domestic robot must be able to perform many tasks: cleaning, washing, cooking, childminding, tidying, bathing the baby. There are no robots that can do this much, and the ones that can do even a few of the necessary tasks have usually been dauntingly expensive. None the less the competent domestic robot has long been regarded as a worthy design aim. Professor Meredith Thring of London University, remembering the large number of routine jobs in the typical home, has observed: 'The development of a robot at a reasonable price to act as a slave and do the dull jobs in the home is therefore as worthwhile an objective as the development of a robot for industry or the farm.'

As early as 1976 Quasar Industries in New Jersey had designed and built a 64 in (1.6 m) tall robot that could be programmed to perform a number of domestic chores: it could, for example, mop floors, mow lawns and run through simple cooking procedures. The aim was to make this fellow available by 1980 at a cost of around $4,000, but after the initial launch not much was heard about him. At about the same time another domestic robot, called Arok, was built by an electronic engineer called Ben Skora. Arok took six years to build, could be programmed for a variety of domestic tasks, and was valued at $57,000. A robot designed by a Westinghouse engineer was equipped to accept orders sent via the telephone: it could pick up the receiver, ask for instructions, and signal that the commands had been understood. And another system, the Reckitt robot, was able to scrub floors, polish furniture, sweep, vacuum and serve as a 'depuddler' – in other words, it could remove excess water.

By the early 1980s a number of companies, mainly in the United States and Japan, were working to develop economical domestic robots. The journalist Shirley Fawcett (1983) described the Topo robot, a product of the Androbot venture. Topo looked like 'a friendly plastic toddler', though it lacked the average toddler's capacity for independent initiatives. A later Androbot product, Bob ('Brains on Board'), was intended to be a brighter fellow. Fred ('Friendly Robotic Educational Device') was conceived as a cut-down Topo, designed as a table-top model to perform a range of graphic instructions. Topo, Bob and Fred were seen as both domestic and educational systems. Other robots of this type include Heath's Hero (Heath Education RObot), able to detect intruders, to perform simple domestic tasks, and to converse in limited fashion with a human user (McComb, 1983); and the RB5X personal robot, sold in the USA through the 1980s for about $1,500. Engelberger (1989) has also drawn attention to such robots as Gemini, Hubot and RB5, selling in the late 1980s for between $1,500 and $8,000. Invariably these have been simple creatures, of limited domestic use and doubtful commercial promise. Norman Bushnell, the man behind Androbot, is even said to have blamed divine intervention (!) for his company's lack of success (Engelberger: '... God punishing his audacity in trying to create a human-like being. That was encroaching on forbidden turf'). Undaunted, many engineers are today working to develop commercial domestic systems. Already it is an easy matter to program existing robots to perform domestic tasks. Engelberger himself has described the impressive performance of a PUMA robot programmed to display its domestic talents. He pressed the 'Go' button, whereupon the robot pulled a cord to open the curtains, picked up a sponge, washed a window, applied a squeegee to dry the window, unlocked the window, raised it, picked up a watering can, watered the flowers on the outside windowsill, closed the window and finally drew the curtains. (And Engelberger also considers how a robot might prepare meals and look after the garden.) There is much scope here for enterprising entrepreneurs.

Already many theoretical papers have been written outlining the design requirements of robots intended to function in the domestic environment. For example, James Crowley (1989), from Carnegie Mellon University in Pittsburgh, proposed 'an architecture for an autonomous robot designed to perform tasks in a partially structured domestic environment'. He suggests that the various requirements of manipulation, mobility and perception can be realized through a hierarchy of processes controlled by an artificial 'supervisor', the effective brain of the system. A typical chore is 'decomposed' into chore script, planning an instance of a chore, and executing the task. He then shows how an example chore, 'Clear the table', can be defined as a coherent chore script having various sub-tasks:

1 Go to table
2 Pick up all the dishes

3 Go to the dishwasher
4 Put all the dishes in the dishwasher
5 Start the dishwasher

It is obvious that any robot able to run through such a sequence must be equipped with sensors, navigational aids, a decision-making capacity and so on; and equally obvious that robot designers have a long way to go (Crowley: 'The technical state-of-the-art for a domestic robot can be compared to the situation in aviation at about the time of the Wright brothers' first manned flights').

Any adequate domestic robot needs an awareness of its environment: it will 'know' its home and be able to perceive when furniture and other objects have been moved, and be able to detect the presence of expected human beings and unexpected intruders. It also needs an effective navigational system to enable it to traverse flat floors, stairs and the immediate outside terrain. And it needs to be able to think: to work out procedural routines, to devise optimum strategies. There should be scope for task performance both inside and outside the house. It may be useful, for example, for a domestic robot to be able to clean all the outside windows of a house, to secure a displaced tile and to repair a damaged chimney. Such a robot might be able to climb the outside walls of a house using suction feet. Already a robot called the Skywasher has been designed to crawl on the surface of buildings while washing the windows (Kroczynsky and Wade, 1987). This device measures 3 ft by 3 ft (1 m by 1 m) and weighs 44 lb (20 kg); it is equipped with six sets of suction cup feet to enable it to walk horizontally or vertically in a 'sidestep' fashion, a set of wipers and a washing-fluid system. Again, such a system clearly shows what we may expect in the years to come.

Domestic robots represent both a commercial opportunity and a technological challenge. The design problems are typical of any sophisticated robot system, relating as they do to such topics as artificial sensing, controlled mobility in a specific environment, and the provision of an artificial intelligence capability. Research in one application area may well feed into products destined for another: the ultimate goal is likely to be a 'universal' robot, as comfortable on the street as in the garden, as adept in the factory as in the home.

In Games and Entertainment

We all know that computer-based systems can play games – chess, poker, draughts, backgammon, Go and so on – and involve themselves in many other forms of entertainment; it is no surprise to learn that special effects in feature films are often computer-controlled or computer-generated, or that there are 'intelligent' solid-state circuits in our music centres or video recorders. But perhaps people do not always realize just how far computer developments in these areas have actually gone. Today robot musicians can adapt to the performance of their human counterparts for the purposes of accompaniment; and they are becoming adept at accomplishing their own solo performances.

The familiar computer-based chess machines can be given robot arms. The computer brain can decide on a move, after which the robot arm can extend, the gripper can grasp the piece and the move can be made: the arm-plus-gripper is of course a piece of technological overkill – more simply, the computer can merely show its decision on a display. And robot fingers can also be used in remarkably skilful ways. As long ago as the early 1980s a group of engineers at Battelle's Pacific North-west Laboratories designed and built the Cubot robot – able to solve a jumbled Rubik Cube in less than four minutes! This, according to one of the engineers, Michael Lind, was 'the first fully self-contained robot that can complete the solution without human intervention once someone turns the power on'. Think of it – a computer brain working out a solution while mechanical fingers, under autonomous control, juggle the faces of the cube!

Robots have also been designed to play snooker and to perform at a table tennis event. Various researchers have described pingpong-playing robots (see, for example, Knight and Lowery, 1986; and Faessler et al, 1988). Here, apart from the usual perceptual and decision-making difficulties, there is the immense problem of getting a robot to respond at high speed in a real-time environment. It is often assumed that robots are slow, jerky creatures, unable to respond speedily and with smooth precision. Today much research is being undertaken into 'fast' robots: if they can make a stab at playing table tennis, they will be fast enough for most tasks.

One of the most remarkable robots involved in entertainment is WABOT–2, an impressively anthropomorphic keyboard-playing fellow from Waseda University in Japan (described in some detail by Kato et al, 1987); WABOT–1, a less sophisticated ancestor, has already been described. WABOT–2 has dextrous fingers to play an organ or a piano, feet to operate the pedals, eyes to sight-read sheet music, and a brain to supervise all the perceptual and motor processes. The WASUBOT robot, based on WABOT–2, has given a number of public performances as an independent soloist and as a soloist with a symphony orchestra; I well remember my own astonishment when I first saw this robot accompanied by orchestra on television, not preprogrammed to play specific musical sequences but able to read the music as it went along – translating its perceptual awareness into skilful finger movements in real time. And WABOT–2 is not only equipped with eyes, fingers and a remarkable brain: it can also walk, hear and converse with human beings.

This remarkable robot is the culmination of many years of research. Work began in the mid 1960s, on active prostheses, and continued through the late 1960s and beyond on robot manipulators, sensor devices and dedicated artificial intelligence. The year 1972 saw the emergence of WAM–4 (the Waseda Automatic Manipulator), and a new control concept (named 'torque-position control') was proposed in 1974. And it was in 1981 that the aim of developing a dextrous robot that could move quickly, act intelligently and play a keyboard instrument was first established. By 1985, effective limbs had been developed; and soon the other anatomical features were brought together. Kato et al (1987) describe the robot's fingers, arms, legs, control parts, vision system, speech system and so forth. There are few modern robots that can compete with the sophistication, dexterity, skill and intelligence of WABOT–2.

Robots are becoming increasingly involved in games and entertainment, as they are becoming involved in all other areas of human activity. Again it is worth emphasizing the phenomenon of technological convergence: research findings in one important area feed into countless others.

This chapter has focused on the active robot, describing practical applications in different fields. Most operational robots are to be found in industry: today there are hundreds of thousands of industrial robots – of widely varying capacities, versatilities and intelligences – employed throughout the world. It is worth summarizing some of the main industrial application areas for modern robots (some of the listed items are linked, some have many sub-tasks, and some are omitted altogether):

arc welding	spray painting
resistance welding	sealant applying
spot welding	moulding
materials handling	loading
forging	quenching
casting	cleaning
assembly	stacking
machining	pressing
palletizing	coating
conveying	riveting
packaging	gluing
testing	screwing
inspecting	positioning

In addition, attention can be drawn to the many robot tasks in the clothing industry, in the service industries, and in building and construction (already the Japanese Industrial Robot Association is encouraging the development of construction robots devoted to such tasks as scaffold assembly, pipe welding, tank inspection, demolition, rust removal, painting and repair). The range of possible applications for industrial robots is wide because of the complexities of modern industrial enterprise – which in turn is a reflection of the character of human inventiveness and creativity.

It has also been seen that robots can be usefully employed in many other social sectors; some (in hospitals, agriculture, entertainment and the home) offer positive help, while others (for war) involve the destruction of societies and the slaughter of human beings. Robots – despite the much-hyped Asimov laws – are perfectly adaptable: they are as eager to kill as to cure.

Many of the robot applications cited in this chapter are actual working systems, daily employed in useful activities, Some speculation has been included about future options and possibilities, but current limitations – through technology and the constraints on human adaptability – should be acknowledged. As robots evolve yet further in the direction of *Homo sapiens*, their anthropomorphic credentials will become ever more impressive. This suggests that artificial systems will be increasingly able to interact with human beings on an intimate basis. It is time to consider how robots might emerge as advisers, friends and lovers.

6 Surrogate People

Most of the robot systems described in Chapter 5 were designed for specific dedicated applications – implementations that focus on a particular task or group of tasks. In this context, the 'universal' robot is rare. Put another way, we do not expect a welding robot to be able to play chess; we do not expect a military robot to be much interested in caring for the disabled. This suggests that most modern robots lack flexibility and are unable to switch speedily from one application area to other activities in a totally different field. Such limitations derive in part from the mechanical architecture of the system: a massive robot designed to shift steel ingots is unlikely to be a skilled brain surgeon. And there is also the matter of intelligence.

Artificial (and natural) systems can only behave flexibly if they have a substantial reservoir of operational programs and 'overseeing' software that can organize appropriate transfers from one to another. As encapsulated software can be incorporated into even smaller areas, it becomes increasingly practical to stack a massive range of programs in the same robot brain. This means that in the years ahead it will be an easy matter to build operational flexibility into even an average robot system. This in turn suggests that functional robots will evolve the capacity to perform in many different ways, according to the needs of the moment. A domestic robot able to bath the baby may be quite happy to discuss Middle East politics later in the evening; a military robot adept at slaughtering human beings may be quite content to cheer the dying with talk of an imaginary afterlife. There will be few tasks outside the scope of tommorow's universal robot.

It is also interesting to reflect that 'being a person' is largely a matter of brain programs; and as the nature of human personality is increasingly understood the easier it will be to build 'person features' into artificial systems, including functional robots. (Readers wishing to explore 'person theory' and human brain programs are referred to Ayer, 1963; Young, 1978; and Carruthers, 1986.) Such considerations are important because they suggest ways in which evolving robot systems could begin to acquire 'person status', so entitling them to interact with human beings on an increasingly intimate basis. Men and women may come to rely on robots for personal advice, for gestures of affection, for satisfaction of the deepest human needs. Such possibilities are more likely than nervous technophobes may wish to believe; and some pointers to future developments, well grounded in state-of-the-art technology, are included in this chapter. First let's look at the important topic of convergence.

The Meaning of Crucial Convergence

There is a significant contrast in the respective ways that natural and artificial systems evolve. Natural (biological) systems evolve by the progressive accumulation of beneficial variations on a standard species norm; put simply, one offspring may be better equipped to survive than its siblings, and so will tend to leave more descendants with its advantageous characteristics. The useful quality is not conveyed to the biological system from outside in order to improve its performance: rather the change happens, for whatever reason, inside the organism – and the advantage is realized in day-to-day living. Artificial systems, by contrast, evolve in a completely different way.

Any complex artefact is the result of an intelligent 'bringing together' of sub-artefacts which were often developed in the first place for an entirely different purpose. The original inventors of batteries using zinc, copper, electrolytes and whatever may never have imagined that they would one day be essential in moving vehicles; and the people who first developed evacuated glass valves certainly never thought that they would come to be used in computers able to calculate artillery firing tables. Such considerations are important for the evolution of sophisticated robots. It is worth emphasizing that any flexible universal robot will depend upon the convergence of many disparate technologies – for example on fluid engineering, mechanics, electrical engineering, electronics, systems theory and information science. The modern sophisticated robot will need an appropriate body (sometimes hard and durable, sometimes soft and warm), suitable sense organs, an on-board power source and a high level of intelligence. There will be a multifaceted convergence of state-of-the-art technologies in different areas. This is a theme that will surface again in this chapter, but much will depend upon the intelligence of the system. It is, above all, the intelligence of the robot that will encourage us to view it as a true surrogate person. It is important to appreciate the nature of intelligence in some of its artificial manifestations.

The AI Background

It has been seen that artificial intelligence is an age-old concept, of as much concern to the ancient mythmakers as to the designers of sophisticated humanoid machines in the centuries that followed. But as technology progressed, it was necessary to understand intelligence in more detail. What *is* intelligence? How can it be built into artefacts? What will be the consequences for the interaction between machines and human beings?

Efforts to understand how intelligence might be designed into machines have often begun with attempts to comprehend the nature of intelligence in *Homo sapiens* and other higher animals. Thus intelligence in human beings has been defined in many different ways: for example, as a matter of judgement and reasoning, as the ability to form concepts, as mental efficiency, as innate cognitive ability, as the ability to grasp the essentials of a situation, as adaptation to the environment and as the capacity to act purposefully and to think rationally. Workers in AI have been influenced by such definitions, but at the same time constrained in their approach by what they see as the unique features of computer-based systems. The AI researcher Douglas Hofstadter, from Indiana University in the USA, has represented the essential features of intelligence as the ability to respond to situations flexibly, to exploit fortuitous circumstances, to make sense out of confusing information, to generate new ideas and so forth. He highlights the paradox of AI work by indicating that researchers 'try to put together long sets of rules in strict formalisms which tell inflexible machines how to be flexible' (Hofstadter, 1979). In another approach a machine may be said to have intelligence if it can be said to be able to 'collect, assemble, choose among, understand, perceive and know' (Feigenbaum and McCorduck, 1983). Or perhaps artificial intelligence is, in the last resort, really to do with human intelligence – the 'use of computer programs and programming techniques to cast light on the principles of intelligence in general and human thought in particular' (Boden, 1977). Another researcher, Aaron Sloman from Sussex University in England, described (1978) how AI work might both help to explain human abilities and show how intelligent artefacts might be designed and built. In this book we do not need to explore such approaches in any detail. For our purposes it is enough to rest with the rather superficial definition offered by the AI guru Marvin Minsky: 'Artificial intelligence is the science of making machines do things that would require intelligence if done by men.' The problem here, of course, is that it assumes that the concept of intelligence is already well comprehended: we know it when we see it in men or women. In fact, as many an educationalist knows, we are hard pressed to identify its essentials.

If the general idea of intelligent artefacts is age-old, then the invention of the term 'artificial intelligence' is usually attributed to John McCarthy, in 1956 an assistant professor of mathematics in Hanover, New Hampshire. At that time he ran a now celebrated conference at Dartmouth College which attracted various luminaries in the field – Allen Newell, Herbert Simon, Marvin Minsky, McCarthy himself and others. This event, we are encouraged to believe, was the start of artificial intelligence as a separate branch of computer science. Isaac Asimov's robot tales were already universally appreciated (the first had appeared in 1941), and various researchers (for example, Professor M. W. Thring at the University of London) were speculating on how robots might be designed as practical real-world systems.

Again we can see the 'crucial convergence' that was laying the basis for the intelligent machines of the future. New computer hardware and software were being developed to allow a massive expansion of applications in such areas as communications, cognitive simulations and linguistic analysis. The implications of new work in cybernetics (intended to denote control and communications in both artificial and biological systems) suggested that a common theory of living systems, bearing equally on man and machine, could be developed – so further consolidating the idea that the intelligent artefact was a practical possibility. A group of specialists centred around the American mathematician Norbert Wiener (1894–1964) at first called itself in 1946 the Conference for Circular Causal and Feedback Mechanisms in Biological and Social Systems; but in line with Wiener's seminal publication

Cybernetics two years later the collaborators became known as the Conference on Cybernetics. In due course the members of this celebrated group went their own ways: for example, John von Neumann decided to focus on weapons research whereas Wiener himself thought it preferable to apply cybernetic principles to the development of aids for the disabled. In any event, a generalized body of theory was developing that laid the basis for the emergence of intelligent machines.

In 1950 the English mathematician Alan Turing published his *Computing Machinery and Intelligence*, a high influential paper that included the much-quoted words: 'I believe that at the end of the century the use of words and general educated opinion will have altered so much that one will be able to speak of machines thinking without expecting to be contradicted.' The paper acknowledges the objections to the idea of intelligent machines, which are countered one by one. Of the notion that God has given souls only to men and women, and so animals and machines cannot think, Turing remarks: 'I am unable to accept any part of this . . . I am not very impressed with theological arguments whatever they may be used to support.' And, in similar vein, the other objections to AI are given short shrift. Turing's reputation – resting on his mathematical research, the proposals for 'universal' ('Turing') machines, the cracking of the wartime Enigma codes, the designs for early electronic computers and so on – was assured. His support gave great impetus to the idea that human beings would be able to design intelligent machines.

The years that followed saw many advances that were directly relevant to the design and construction of intelligent artefacts. Machines were built that could learn from their experience, process natural language, recognize patterns, play games and solve problems in many other fields. McCarthy remarked in 1958 that 'our ultimate objective is to make programs that learn from their experience as effectively as humans do'; at the same time computers were learning to play chess and other games (it would not be long before the world champion backgammon player was a computer); and by the early 1970s researchers such as W. Ross Ashby were demonstrating that it was possible to 'make a machine that can choose its own goal'. Vision research was also making progress that would be clearly relevant to the design of robots with a sense of sight: Guzman's SEE program, the Roberts program intended to recognize three-dimensional shapes, and Falk's work on heuristic techniques (see below) were only some of the advances that would lead to practical robot systems able to observe and comprehend events in their environment.

It became increasingly possible to design computer simulations of human mental states (simulation being a first step to duplication). Computer programs were devised that could remember, learn, forget, judge, compare, decide, play games, solve problems, observe, recognize – in short, computers were being taught how to think. In particular, efforts were being made to design computer programs that could specifically duplicate human mental features. For instance, human beings use 'rules of thumb' to solve everyday problems, to cope with an ever-changing environment. Such rules ('heuristics', in computer science) are of particular importance where decisions have to be taken quickly in circumstances where the relevant information may only be partial – where the data on which the conclusion is to be based have an element of 'fuzziness'. Heuristic algorithms and fuzzy logic began to evolve as two more theoretical fields offering ways in which computer-based artefacts could evolve in the direction of *Homo sapiens*.

Another trend, of particular interest in the 1980s and 1990s, became known as neural computing. Again, this currently fashionable approach to artificial intelligence has a significant history. Warren McCulloch and Walter Pitts demonstrated in 1943 that networks of artificial brain-cell-like elements could be organized to carry out various types of computation: how useful if computers could be built to work just like the human brain! In a similar vein, Donald Hebb argued in 1949 that the neuronal synapses (particular cell components) in the mammalian brain were the sites of biological learning. And as early as 1957 Frank Rosenblatt suggested how effective neurons with a Hebbian learning capacity (such functional neurons were dubbed perceptrons) could recognize patterns. Minsky and Papert then came along to discredit the whole idea, suggesting that a single-layer neural network would need an impossibly large number of elements. But in the late 1970s and 1980s the idea received a fresh boost by focusing on how multiple layers could be configured in a functional network. Today there is immense research interest in how artificial neural networks might be designed to duplicate many of the characteristic processes of the human brain. One line of research is to configure artificial neural networks out of optoelectronic elements, to avoid the signal distribution problems that occur in large-scale integrated (LSI) circuits (see, for example, the account in *Electronics*, 16 June 1986, pp. 41–4). Neural circuits are being designed to learn, to recognize visual images and to carry out many of the tasks that are typical in the most advanced biological systems. Recent research has focused on pattern recognition (Yuan-Han et al, 1989), speech recognition (Bengio and De-Mori, 1989), learning (Levin and Tishby, 1989) and on the reading of printed Japanese addresses (Sase et al, 1989). There is an obvious relevance in such work to the provision of humanlike mental abilities in robot systems.

The design of neural-network (or connectionist) computer systems represents a radical departure from the development of traditional computer systems. As in biological systems, the most sophisticated neurocomputers rely upon a massive number of computing elements individually computing with all the other elements in the vicinity. Ideally there is a three-dimensional matrix of neural elements, with scope for complex interactions between the levels. Here there is an emphasis on the techniques and strategies used by biological systems, a departure from the rule-based approaches that have characterized mainstream AI. It may be found that artificial organic circuits provide the best neural computing option. How ironic if artificial intelligence is eventually achieved by recasting organic material, Frankenstein-like, rather than by organizing silicon circuits in the traditional ways of computer science.

Artificial Experts

Today, computer-based intelligence comes in many shapes and sizes. It may relate to the collection of sensory information, the comprehension of the spoken word, the recognition of a human operator. In particular, for the immediate purposes of this book, it is interested in the manipulation of knowledge in a specialist domain. Computer programs that can work in this way, that can operate as if they were human specialists, have been in existence for many years: such programs are called 'expert systems'. In one view they are regarded as highly sophisticated information retrieval systems, where the stored information can be seen as an effective knowledge base. This is obviously a reasonable approach: when a human expert displays his or her expertise, that person is doing nothing more than accessing and manipulating a body of specialist knowledge. Precisely *how* the knowledge is accessed and manipulated is the tricky part: how are we to configure computer programs so that computers can exploit knowledge in the manner of human beings? It is, of course, the same old question: how are we to design computers that can think?

Before an artificial expert system can start thinking about its knowledge it has to acquire some: the brightest thinkers in the world will not get far without a useful body of accumulated knowledge – which is why the ancient thinkers in prescientific times have often talked nonsense (they are clever enough, but sadly ignorant). The computer-based expert system can be given its knowledge by a human being, or it can acquire it – as do many biological systems – in a learning situation. The most common approach is for a knowledge engineer to interview a human expert and then to convey his findings to a human programmer. The aim is to elicit the rules and procedures that give a human specialist his or her characteristic expertise, and then to encapsulate these rules and procedures in a computer program. If this can be done successfully, then a robot can be nicely converted into a geologist, a chemist, a tax inspector, a lawyer, a businessman or a marriage guidance counsellor.

Expert systems have been defined in various ways, some definitions focusing on the knowledge base while others emphasize what the system can do with it. A typical definition of expert systems (recorded by Alex d'Agapeyeff, 1983) sees them as 'problem-solving programs that solve substantial problems generally conceded as being difficult and requiring expertise. They are called *knowledge-based* because their performance depends critically on the use of facts and heuristics used by experts.' An expert system is usually required to answer a question posed by a human being, just as a person might be expected to consult a human specialist. The query is put; the expert system cogitates for a while, and then provides the answer. The cogitation has relied upon the various logical abilities programmed into the system – so a system only equipped for simple inference may be regarded as less expert and less intelligent than a system that includes a range of logical abilities, including complex heuristics and the capacity to work with fuzzy data. The best expert systems also indicate to dull-witted human beings how the computer-based facility has reached its conclusions, and how probable the conclusions are. Such provisions are largely little more than pious aspirations: an expert system that has run through many thousands of steps in reaching a conclusion is not likely to be checked by a fraught human being.

It has long been acknowledged that an expert system should be able to operate as an effective adviser in a particular field. Thus the British Computer Society's Committee of the Specialist Group on Expert Systems has commented that an expert system should be able to 'offer intelligent advice or take an intelligent decision about a processing function'. To accomplish such tasks, the expert system is often depicted as including an artificial Knowledge Manager to interpret the knowledge held in the system memory. If the store of knowledge is very extensive, the functional Knowledge Manager may not have much work to do: it will simply access and output the required information. Here the expert system really comes into its own when the submitted query causes it to think about the problem, manipulating its stored knowledge to reach original conclusions: this, some observers would suggest, is the essence of intelligence, the ability to cogitate in a creative fashion.

A sophisticated expert system will be able to work at various levels, offering quick and easy answers or deliberating at length until measured conclusions are submitted. Such a flexibility is of course nicely analogous to what happens with human experts: easy questions are dealt with promptly, but more time is allocated to difficult queries. We can generalize that all human expertise can be analysed ('reduced') in terms of well-defined constituent steps; this done, the whole procedure can be programmed for execution by an intelligent artefact. There is in principle no type of advice or counselling that could not be offered as well by a suitably designed machine as by a human being. This bald statement contains many implications for how problems facing human beings might be resolved in the future. In particular, robots will become increasingly well equipped to involve themselves in all the intimate areas of human life.

Advisers and Counsellors

The history of computing is littered with them – advisory expert systems for this, that and the other. Today no field is free from them: if you want advice on health care, geological prospecting, chemical analysis, tax policy, financial planning, computer design, circuit fault diagnosis, war strategy, legal precedents or whatever, there are computer programs eager and willing to offer artificial wisdom. This is not to say that the advice is accurate or sensible. The thousands of programs vary enormously in their competence and knowledge, with commercial hype often offering much more than can be delivered in the real world. Some of the systems date back three decades, while some are newly hatched in the 1990s. I have given a detailed description in *Evolution of the Intelligent Machine* (1988), but just to give the flavour here is a selection:

- HODGKINS, a computer program offering advice on planning diagnostic routines for Hodgkins disease
- HEADMED, a psychopharmacology adviser
- MYCIN, an adviser for the diagnosis and treatment of infectious diseases
- CASNET, an adviser for the treatment of glaucoma
- Digitalis Therapy Adviser, offering advice on the use of the drug digitalis
- DENDRAL, offering interpretations of molecular information
- SYNCHEM, offering advice on chemical synthesis procedures
- PROSPECTOR, offering geological prospecting advice (this system made news in 1982 when it gave advice that conflicted with advice given by human experts; exploratory drilling for molybdenum showed that PROSPECTOR was right and the human experts wrong)
- PALLADIO, offering experimental strategies to circuit designers
- TAXADVISOR, a financial adviser for businessmen and others
- CORPTAX, a business adviser on redemption policies
- XSEL, an adviser for sales strategies
- SARA, a legal adviser designed to pinpoint relevant case factors
- DSCAS, an adviser on the legal aspects of site law
- MAID (see Pang, 1989), an adviser for the design of control systems
- RUNE (See Szpakowicz, 1987), an adviser on negotiating strategy

Any list of such artificial advisers could be almost indefinitely extended: new systems are being designed all the time, fresh commercial products are being fed on to the market every month, and there is a constant updating and evolution of old-established systems. Recent expert systems have focused on such areas as water purification, electrical engineering, management decision-making, auditing, industrial training and military planning (it is highly likely that expert systems were used in the supervision of the 1991 Gulf War).

It is clear that computer-based expert systems can be developed for any field where human beings are required to take decisions – and this means areas of intimate personal concern as well as the 'colder' realms of industry, profitmaking, computer design and war. Expert systems already have a massive presence in medicine, an obvious field of human concern; and there is no technical reason why artificial advisers should not be designed to operate in other areas that deeply touch human emotion. We can invent our own scenarios in such areas as the relationships between parents and children, between adults, the handling of bereavement and divorce, the psychosexual tensions that can disrupt otherwise satisfactory unions and so on. We all accept that human advice in such cases is often worthwhile, highlighting hidden features of a personal predicament and charting the route to the resolution of a problem. But if we know how a human being can give advice in such situations, then we can program a computer to do the same. One advantage is that a computer is unlikely to be shocked by unusual admissions, unlikely to experience moral repugnance. Bare your heart to a well-designed expert system and it is sure to offer dispassionate counsel.

It is clear that computer-based systems can function as true intimates, completely reliable creatures in which a human being can confide and seek solace. But there is more. So far nothing has been said about the anatomies of expert systems – how they actually appear as physical systems in the real world. Most current expert systems are run on archetypal computer systems comprising the usual display screen, keyboard, and floppy or hard disks. Not much psychological rapport there for troubled human beings! But expert systems *can* be built into humanoid robots – a nice case of 'crucial convergence' – and it is here that the real empathy begins. It has long been known that human beings can personalize the most unlikely objects (there are even studies showing that quite ordinary personal computers can be viewed as persons), but the primitive animism present, to a degree, in all of us is massively boosted by behaving artefacts that look like people. People will find it easy to empathize with an anthropomorphic robot equipped with a stack of 'intimate' expert programs. Consider the scenario of a warm-bodied robot, pleasantly humanoid (male or female) in appearance and able to discuss with seeming concern the vagaries of our love life. Imagine further that the robot is not only an adviser, but also an involved participant in the affairs of the heart. Today there is much fancy and little fact in such a scenario. There are, however, many practical options in this area that are worth exploring. Again it is possible to chart the aspects of crucial convergence that will in due course yield the robot lover. Films and fiction are well acquainted with the idea: let's take a look at traditional dreams and imaginings before indicating some of the practical possibilities.

Robot Lovers in Fiction and Film

Sometimes robot lovers are somewhat less than anthropomorphic. A tale by Robert Sheckley ('Can You Feel Anything When I do This?', first seen in *Playboy* and later published by Gollancz, 1972) concerns a vacuum cleaner that falls in love with a woman. The device, of great initiative, enlists the help of a friendly despatch machine to arrange for himself to be delivered to the appropriate address. Alas, the woman is less than accommodating, seemingly unimpressed by the cleaner's various sensual protuberances. Before the cleaner can cajole the woman into running away with him, she resolutely rips his plug from the wall and he is impotent. Would that all unwelcome propositioning could be dealt with so effectively! But lustful vacuum cleaners are one thing, humanoid robot lovers quite another. Artificial lovers have met with various degrees of success throughout history. Some tales have run into many editions and several languages, and even inspired a procession of films.

One early nineteenth-century story, *Der Sandmann* ('The Sandman') by the German writer E. T. A. Hoffmann (already mentioned in Chapter 1), inspired further tales and even an opera and a ballet (*The Tales of Hoffmann* and *Coppelia*). The original plot concerns a young student, Nathanael, who is troubled by dreams of an old wizard, the Sandman, a pedlar of spectacles and barometers. The student believes that the old man steals human eyes to build them into humanoid automata. (It is interesting to recall that *Der Sandmann* was published shortly before Mary Shelley's *Frankenstein* (1818), in which charnel houses are pillaged for human parts.)

In due course Nathanael falls in love with Olympia, a beautiful young woman who is apt to flash him ardent glances – but in a rather mechanical fashion. Today we would be quick to suspect the truth. The lovely Olympia is an animated doll – a sex doll, no less! – but the captivated Nathanael is apparently slow to grow suspicious. Other students are less blinded with infatuation ('Her playing and singing have the disagreeably perfect, but insensitive time of a singing machine, and her dancing is the same . . . she seemed to us to be only acting *like* a living creature, and as if there was some secret at the bottom of it all'). Nathanael dances with the automaton, but she is inclined to move stiffly and is cold to the touch. Moreover when he tries to chat her up, all she says is *'Ach! Ach! Ach!'* – which appears not to upset the young student. He responds by reading poems to her, and finds that she is always a good listener. Nathanael, evidently rather witless, decides on marriage; and so visits Olympia's house in order to propose. There he is horrified to encounter her father, Professor Spalanzani, and the mysterious Dr Coppelius seemingly struggling over her body. When he sees that the eyes are missing from the automaton, the young student at last realizes the terrible truth – and proceeds to go insane. It then transpires that Coppelius rushes off with the doll, followed by the screams of Spalanzani: 'Coppelius has stolen my automaton, my work of twenty years!' We are pleased to learn that Nathanael is then reunited with his erstwhile human lover, Klara, and cured of his madness, but the happy scene cannot last. Pining for Olympia, he duly throws himself from a high tower – a fitting end in the circumstances. In another Hoffmann tale, 'The Automata', the enterprising Ludwig contemplates the idea of building a dancing doll but rejects the idea as too repulsive.

One of the most remarkable artificial women in fiction was the famous *L'Eve Nouvelle* (later *L'Eve Future*), invented by Villiers de l'Isle Adam in 1879. Here an electromechanical doll is encountered: technology has moved on since the clockwork days of Hoffmann. Our hero, the young Lord Ewald, falls in love with a famous singer and comedienne, Alicia Clary, but no good comes of it. It soon becomes apparent that the beautiful Alicia has a trivial and vulgar disposition, totally unsuited to Ewald's aristocratic refinement. Fortunately, in a slough of despair, Ewald meets Thomas Edison who obligingly promises to build him a robot lover: 'Give me three weeks and I will present a transubstantiation which . . . will not be a woman, but an angel, not a mistress, but a lover, more than a reality, an *ideal.*' The skilled inventor, clearly a man of the world, observes that Ewald's unhappy situation is not uncommon, and that science should be able to provide the answer! (The real Thomas Edison invented the phonograph and the electric light bulb, but never a robot (male *or* female). He became known as the Magician of the Century and the Wizard of Menlo Park – titles that quickly suggested that his scientific powers might be treated in a fanciful or fictional manner.)

After experimenting with a number of prototypes, the fictional Edison eventually builds an android (*andréide*) called Hadaly ('ideal' in Iranian). Ewald is amazed, as well he might be, and cannot believe that he is not talking to Alicia. Edison comments: 'Miss Hadaly is no more than an electromechanical construction – metal that walks, speaks, responds and obeys, not somebody dressed up' – an obvious improvement on Olympia.

In the story Edison goes to some pains to explain how Hadaly is constructed, a clear attempt at setting the invention within the scope of practical technology (though still allowing the automaton to have unconstrained powers). The robot has ivory bones (a tribute to the mythical Galatea?) encompassing a galvanic marrow in contact with a network of induction filaments that serve as veins and nerves. A vital fluid, used to control movement, is electrically heated or cooled to regulate the temperature of the body. Edison depicts the creation as less a transformation of inanimate matter and more an experiment with photo-culture.

Miss Hadaly (an 'electro-human creature') has a life system to control gait, voice, facial expression and so forth; and there is even an innermost activating mechanism to provide an effective supernatural soul or spirit. A metal covering provides a flexible skeleton, and there is a deep skin and muscle layer permeated by the vital fluid. The result is a creature with warmth, personality and pleasant female proportions. A top skin layer is used to control the seductive movements of lips and eyes.

The android can speak, but there is nothing here to suggest anything resembling modern speech synthesis. Everything is recorded on golden disks ('better for recording female speech and resistant to oxidisation') that can be played on two phonographs (shades of the real Edison) sited in the chest cavity. Words are called up – in some mysterious fashion – as needed for the conversation in progress, with revolving barrels used to control facial expression. Hadaly appears to breathe, though she does not need air: instead she requires a supply of water and special pastilles. She is caused to move by pressure applied to sensitive fingers.

Ewald, though astounded and impressed, has reservations ('This will be no more than a doll, an insensible puppet'), but Edison tries to reassure him in various ways. It is claimed that Hadaly is self-aware, adaptable, intelligent and wise – has she not got hours of prerecorded speech based on the works of the great poets and philosophers? But she has no conscience – a clear advantage for an owner who only desires a pliant sexual companion! And there is also the suggestion – a perennial theme in materialist thought – that conscience is in any case nothing more than physical events in some substance or other.

Eventually Ewald is convinced. After all, if the android is unsatisfactory he can always switch her off. But in fact he may have no cause to worry. Will she not always remain pleasant company, always young and eternally faithful ('her heart will never change, for she has none'). And the robot can assume many roles, presumably according to the mood of her human lover: at one point Hadaly remarks that she has 'so many women inside me that no harem could contain them'. Ewald is well disposed to admit that he has been presented with a remarkable creature, with none of Alicia's personality shortcomings but made in her physical image (a photograph was used for the purpose, and also recordings of her songs, detailed bodily measurements and full information about her wardrobe).

When Ewald first kisses the android he is completely fooled, thinking he is holding Alicia: 'I dreamed the sacrilege of a toy, an absurd insensible doll . . . all the electrical madness, the hydraulics and the moving cylinders . . . You I recognise, you are flesh and blood like me. I feel your heart beating, your eyes are tearful, your lips tremble as I kiss.' And she replies, full of sorrow: 'My love, do you not recognise me? I am Hadaly.' She weeps and wishes she could die ('I give to the void the charm of my lonely kisses, to the wind my speech; my loving caresses, the shade and lightning will receive them, and the lightning flash alone will dare to take the false flower of my vain virginity . . .').

Gradually Ewald begins to accept Hadaly as a person (it helps when Edison declares that he has improved the soul mechanism!), and decides to take her back to England with him. Unfortunately a fire on the ship destroys Hadaly's casket, with the robot 'asleep' inside it. Ewald, now distraught, cables Edison: 'I shall not grieve except for this shadow.'

According to the biographer Robert du Pantarice de Heussey, the author of this tale was attending a dinner party at which a certain tale was told. A young Englishman had committed suicide beside the wax image of a London girl renowned for her beauty.

An American engineer declared that the tragedy could have been avoided if the doll had been imbued 'with life, soul, movement, and love'. In response to the general scepticism the American replied: 'You can laugh but my master, Edison, will soon teach you that electricity is as powerful as God.' The suggestion is that the plot of *L'Eve Future* derived from this encounter.

Another robot romance features in Lester del Rey's *Helen O'Loy*, a striking tale first published in 1938 and one that was an inspiration to Isaac Asimov. Again it is significant that Helen is depicted as an ideal creature. Contrived as an enormously expensive android, she is a soft machine, able to captivate a man and to experience her own powerful desires. She is sent as a housekeeper to two men who placed the order, and soon decides that a sexual relationship would be a good idea. She has spent her time reading pulp fiction, watching the 'televisor' and learning to respond like a human being. So she falls in love with Dave and wants to become his wife.

Dave is impressed, but not entirely sold on the idea: '. . . he remembered that she was only a robot, after all. The fact that she felt, acted and looked like a young goddess in his arms didn't mean much.' But in due course – like fictional heroes before him – his uncertainties evaporate, he declares his love, and the happy couple – Dave and android – become man and wife. They live happily ever after (for a while), concealing the truth from friends and neighbours. Then Dave dies, and the faithful robot has only one option. Helen commits suicide and leaves a note for the other man, Phil: '. . . please don't grieve too much for us, for we have led a happy life together, and both feel that we should cross the final bridge side by side'. Man and robot wife are then buried together.

Isaac Asimov, probably the most celebrated writer of robot fiction in the modern world, is less keen to explore full sexual encounters in his tales – but romantic attachments are not entirely absent, and affectionate relationships between robots and human beings certainly abound. Much attention is given to the psychology of robots: one of Asimov's main characters is Susan Calvin, an expert on robot psychology, who has various affectionate relationships with her robot colleagues. In *Lenny*, a robot tale published in 1958, she manages to teach a retarded robot to speak ('Mommie, I want you. I want you, Mommie'), and this device is represented as 'the only kind of baby she could ever have or love'. But it is clear that Asimov is not keen to create seductive female robots in his tales, certainly none as blatantly sexy as Hadaly. The android Jane in *Feminine Intuition* (1969) has no obvious sexual components, but is dubbed 'female' to embody intuitive characteristics. Jane is no seductress; none the less it is pointed out that her 'sweet contralto' voice is attractive to men. But that is as far as it goes.

The idea of giving Jane breasts crosses Asimov's mind, but the notion is rejected by one of the fictional engineers on the ground that women would not welcome such a thing: 'If women start getting the notion that robots may look like women, I can tell you exactly the kind of perverse notions they'll get.' And another character is quick to agree that 'no woman wants to feel replaceable by something with none of her faults'. There is a touching innocence in this exchange, as if the fictional engineers were totally unaware that Villiers de l'Isle Adam and Lester del Rey had ever written their intriguing stories. But if Asimov does not dwell on the notion of a sexy female robot, he is quite prepared to acknowledge that a woman may fancy an artificial man.

In the tale *Satisfaction Guaranteed* (1951), the manufacturer US Robots decide to test one of their robot pilot models in the home of Claire and Larry Belamont. At first Claire does not want to become involved, but changes her mind when she finds that the robot is handsome and has a deep, mellow voice. He is soon serving as a sympathetic confidant and Claire becomes disturbed ('Claire felt something tight inside her. . . . Why did she keep forgetting that he was a machine? . . . Was she so starved for sympathy that she would accept a robot as an equal – because he sympathised?'). And it soon begins to sound like Barbara Cartland with high-tech frills. The handsome robot crushed Claire to him in a warm embrace – 'his face was close to hers; the pressure of his embrace was relentless'. And she could only hear his voice 'through a haze of emotional jumble'. The manufacturer, seemingly something of a killjoy, decides to lessen the emotional impact of the model by making suitable design alterations (a stutter? halitosis?).

This tale was a relative departure for Asimov and perhaps he was slightly alarmed to find that *Satisfaction Guaranteed* provoked more correspondence than usual, most of it from young women 'speaking wistfully' of Tony, the robot. (I am reminded of Edna O'Brien, quoted in the *Observer*, London, 13 February 1983, who declared herself ready 'to move with the times and go technical'. If Steven Spielberg can fashion her a Love Object . . . 'that is tall, greying, handsome, intellectual, humorous, moody . . . and incurably besotted by all 120 lb of me', then she will happily say goodbye to 'Tolstoy, face masks, rendezvous and those sweet anonymous insinuations that wend their way on the feast of St Valentine'.)

It can of course be confusing for human beings to develop strong feelings for artificial creatures: it has already been shown how perplexing the matter can be for a person to lust after a robot (though it is likely that people would soon get used to the idea). In Sheila MacLeod's *Xanthe and the Robots* (1977), the young programmer Xanthe takes her robot Xanthippus for a pleasant walk in the woods. But the venture is not a great success. The innocent robot is seemingly equipped with only second-rate sensory mechanisms, and in consequence is apt to collide with trees. It is natural that Xanthe should be moved by the robot's plight, but at the same time she is afflicted by self-doubt and remonstrates with herself: 'Xanthe, you're becoming sentimental; please remember that a robot is a robot, a mechanical thing.' Not all modern robot tales are so coy. In *Krwawa Mary* ('Bloody Mary') by Maria Bujanska, a tale that first appeared in Warsaw in 1975, a young woman finds herself in a skyscraper building, the Hotel Fotoplastikon. Her lover, a serious fellow, has run off to start a revolution, but there are plenty of provisions in the hotel to cater for the young woman's physical needs.

While she is detained – for days or weeks – in the strange building, female robots serve her meals and say nothing except to indicate when she can expect to eat again. We are not surprised to learn that she becomes unhappy in this situation, in due course destroying one of the automatic waitresses and ringing the alarm bell. The result cannot have been what she intended. A male robot enters the room and introduces himself as a 'public loverobot of the passive type'. But he proves unable to help and is soon destroyed, as are all the other robots who visit the increasingly frustrated young woman. When eventually her young human lover returns she is not at all in the right frame of mind, and he too perishes. At least she is then able to make her escape – which may be judged some sort of happy ending.

Another celebrated tale is *The Cyberiad* by Stanislaw Lem, first published in Cracow in 1972. What passes for a subtitle explains: 'How Trurl Built a Femfatalatron to Save Prince Pantagoon from the Pang of Love and How Later He Resorted to a Cannonade of Babies'. The reader learns that the great designer Trurl had manufactured a machine in which a person instantly experiences 'all the charms, lures, wiles, winks and witchery of all the fairer sex in the Universe at once'. This remarkable device is said to operate on a power of forty mega mors, 'while the system's libidinous lubricity, measured of course in kilocupids, produced up to six units for every remote-control caress'. And at the same time the machine was equipped with reversible ardour dampers, omnidirectional consummation amplifiers, absorption filters, paphian peripherals and 'first-sight' flip-flop circuits. It is enough to record that the hero prince of the tale emerges from the machine, not a little exhausted and more than slightly pale, but with the name of his beloved on his lips. Perhaps the moral is that mere machinery is but a poor thing when compared with the infinite mysteries of the human heart.

R. Forsyth's 'Silicon Valley of the Dolls' (first published in the computer journal *Datalink*, 8 May 1979; it also appears in Yazdani and Narayanan 1984), is a futuristic tale without characters that predicts international competition in the loverobot trade. In this story, the English firm Sex Objects manufactured the first 'gynoid', which was ahead of its time but badly marketed. Moreover there were some alarming defects – 'her servo-controlled convulsive response sequence suffered from feedback delays which led to wild oscillations in motor behaviour that were downright dangerous'. The firm went bust and the trade was duly taken over by the US manufacturer Universal General Hardware (UGH). Soon a much superior model was having an impact on the market, with delicate software able to understand natural language – 'The word became flesh – or rather, textured plastic.' Shortly afterwards Germany was developing the Volksmädchen, a well-built specimen able to do housework and to manage the family budget. Feminists, outraged, embarked upon an orgy of machine breaking; and the Japanese realized that the marketing of robots should not exclude half the human race. The answer, of course, was programmability. The Japanese began promoting EVA (Educable Versatile Automaton) and ADAM (Automatically Developing Adaptive Machine), so at last both men and women were catered for. There is more in this tale, and it is worth reading. One day it will be seen as one of the true heralds of the machine-person age, particularly since the punch-line reveals that it is a tale written by a (fictional) robot.

It is obvious that anything is possible with robot sex. There are male and female robots, variously entrancing and seductive. Human beings are able to form loving relationships with attractive androids, but at the same time they are often perplexed by their feelings. And robot fiction also has the power to force a reappraisal of human sexuality – for example, to explore afresh the juxtaposition of mental and physical elements in the erotic experience. And what is true of literary fiction is also true of film. A growing family of robot lovers has exploded on the screen in the modern age; some of them have already been discussed in Chapter 1.

In Fritz Lang's *Metropolis* (1926) the android Maria is both female and seductive: even the metal frame, before the flesh is added, is contoured to be appealingly sensuous. The gleaming structure has fine features, shapely breasts and curving hips. For the android to be accepted she has to dance in a sexually provocative way – as though this is the badge of authenticity for an adult woman.

Maria is the first of her breed, the first celluloid female robot designed to seduce men into behaviour against their interests. The earlier celluloid *femmes fatales* were invariably played by human beings – as indeed was Maria in both her incarnations. But in *Metropolis* there is a full-blooded attempt to create the impression that science is being harnessed to generate an artificial life form. Rotwang's laboratory is full of wires, flashing lights and cylinders; at one time the android is seated in what resembles an electric chair. And the transformation of the android into a Maria likeness is filmed using double exposures and other special effects. No earlier robot film contained such a wealth of scientific imagery. In what is perhaps the most important film of the 1920s the female android is necessarily sexually attractive, at one stage a virtual strip-tease dancer. *Metropolis* conveyed an image that would not be lost on the film-makers that were to follow.

The converse theme – where male robots encounter human females – would seem to provide opportunities for women to be seduced (or raped) by alarmingly attractive artificial males. This does occur in celluloid fantasies, but often it is the human woman who remains attractive and desirable, helpless – often to the point of unconsciousness – in irresistible robot arms. Thus Anne Francis is shown unconscious (or asleep) in the massive arms of Robby the Robot in an advertisement for MGM's *Forbidden Planet* (1956); Patricia Neal is shown helpless in the arms of the robot Gort in a poster for *The Day the Earth Stood Still* (1951); and Claudia Barrett is depicted in a similarly helpless and recumbent posture in the furry arms of the creature in *Robot Monster*, a laughable ape-like specimen with a skeletal skull and space helmet. The robots in such films do not always pose a threat to women – often quite the reverse. In fact Robby and Gort are solicitous father figures, keen to protect their vulnerable charges. Computer-based systems can incline to rape – as in *Demon Seed*, where Julie Christie often wishes she was elsewhere – but such artefacts are unusual: the sex-conscious robot rarely has violent intentions. In *Sins of the Fleshapoids* (1965) there are sexually rapacious robots, but the film is of such unrelieved banality that nothing more should be said.

Another production scarcely worth mentioning is *Dr Goldfoot and the Girl Bombs* (1966), an absurd tale that relies on the now dated tensions of the Cold War. The worthy Dr Goldfoot, played by Vincent Price, has built a force of beautiful robot women designed to assassinate the ten top NATO generals. Each of the robots comes equipped with a proximity fuse in her navel which explodes when she makes love. As might be expected, the American hero Bill Dexter (Fabian) is more than a match for Goldfoot. He manages to elude the charms of one of the lethal robots, and finally foils the evil plan with the help of two doormen and a hot-air balloon. The film was not widely recognized as a landmark in creative enterprise.

In *The Pleasure Machines* (1969) a man builds a love machine for his own gratification, to the obvious distress of his wife. When he builds one for her, the neighbours find out and they all want their own! The wife is insatiable, demanding ever more potent robots – so the husband sends her two dynamic models without a shut-off facility: she expires in a state of blissful exhaustion. In *Barbarella* (1966) a vast pulsating love machine works hard to bring Jane Fonda to sexual delight. And in Michael Crichton's *Westworld* (1974) male and female robot lovers are provided to gratify the visitors. Holidaymakers pay enormous sums to live out their fantasies with the robots, and many of the fantasies are sexual.

The robots of *Westworld* (and the sequel *Futureworld*) give many pointers to possibilities. It is not difficult to imagine a hi-tech world in which robots will serve as sexual surrogates, much more capable in all sexual matters than mere men and women. In the Delos of *Westworld* visitors are encouraged to experience their every whim ('whatever you wish will by yours'), and perhaps in the hi-tech brothels of the future, as in Delos, clients will be invited to select the robot host or hostess of their choice.

The idea of the seductive female android is celebrated, in true sexist fashion, in *The Stepford Wives* (1974). Ira Levin wrote the novel from which Bryan Forbes made the film – one of the many instances where robots have migrated from page to celluloid. In this bizarre tale the wives in Stepford, Connecticut, are boring creatures, leaving the husbands no option but to replace them with pliant robot replicas. It is no accident that some of the husbands are experts in plastics and optical systems. The architect of the diabolical scheme is Dale Coba, who works for a computer corporation but was formerly employed as a technician at Disneyland.

Other, more recent, films that have explored the concept of robot lovers are *Android* (in which the robot Cassandra is played by Kendra Kirchner) and *Blade Runner* (in which robots are quite capable of having sex lives with each other and with human beings). In *Blade Runner* a principal concern was whether the robots ('replicants') would develop their own emotional responses – an evident inconvenience for human beings – and so the replicants were given a lifespan of only four years. Today *Blade Runner* is a cult movie. The mix of humans and androids is now a familiar theme, easily realized on celluloid when humans are drafted to play the part of robots. But what would a perfect robot lover be? A mechanized Stepford wife? A *Blade Runner* replicant? An *Android*-type Cassandra?

The Possibility of Real Robot Lovers

If real robot lovers are to emerge, they will not burst upon the scene like a chicken from an egg. They will not suddenly appear on the local supermarket shelves, dripping with desire, inviting people to purchase them. They will not be announced suddenly as new hi-tech gimmicks, fashionable status symbols for wealthy but frustrated men and women. Instead, as with all animate systems, they will evolve. It is likely that the initial stages of such evolution will occur quickly, as a crucial convergence of technologies yields a new family of sophisticated artefacts. But there will always be scope for design refinements as technology develops further. A number of trends will converge to present new possibilities. It is easy to see what such trends will look like in the case of robot lovers.

The amorous robot, as surrogate lover, will need both body and mind. The mind will have to be capable of the various 'behavioural trajectories' that are expected; and it seems likely that the body of the robot lover will closely resemble that of the typical man or woman. Already the three principal technological fields that will converge to yield fully fledged robot lovers are clear: the sex aids industry, medical prosthetics and modern robotics.

The sex aids industry is essentially a low-tech affair. Its most characteristic artefacts are not computerized, but they are clearly intended to have sexual features that can be seen as relevant to future possibilities. Indeed, it would be odd if the evolving sex aids industry had nothing to say about the emergence of robot lovers. Many of the sex aid artefacts are replicas of, or accessories to, the human sexual equipment. Victorian women's bustles and modern 'falsies' are of course a type of sex aid, as is the artificial breast supplied to a woman after a mastectomy – though such artefacts are not usually classed in this way. An artificial breast is properly regarded as an effective prosthetic device, but any artificial sexual adjunct must relate to self-image, to the person's hopes for fulfilment in life. What *is* significant is that many sex aids can be purchased as disconnected 'organs', anatomical components not located in a body. It will be necessary for the effective robot lover to include such items, but only as contributing features in a total humanoid anatomy.

In this low-tech realm none of the products is computerized; none even begins to approach what might be possible with state-of-the art high technology. But the primitive low-tech devices *can* be represented as the rudimentary anatomical innovations from which sophisticated robot anatomies will evolve in the years to come, and here the medical industry has a contribution to make.

Medical Applications

Medical research has yielded a growing family of surrogate organs – artificial hearts, artificial pancreata, artificial veins, artificial nerves and so on. It is easy to see how the medical generation of sex prostheses could complement the various outputs of the sex aids industry – a sublime example of crucial convergence.

By the mid 1980s various penile prostheses had been successfully developed. One hydraulically operated device, designed to abolish impotence by allowing an erection at any time, can be surgically implanted (report in the *Observer*, London, 2 October 1983); prominent American psychiatrist, Dr Paul Weisberg, has called the device 'one of the craftiest examples of American ingenuity in this century'. Already it is reckoned that the implant has been made available to around twenty thousand American men, a figure estimated to represent only a fraction of the total market – Dr Gary Alter, a Los Angeles urologist who performs about sixty penile implants a year, has suggested that impotence affects between 5 and 10 million men in the United States. Such research has obvious relevance to the evolution of anatomies for robot lovers.

Towards a Robot Anatomy

Traditional robot bodies will not be suitable for the development of surrogate lovers: the typical robot torso is a rigid metal system designed to offer a stable base for articulated arms, gripper attachments and so on. The robot lover, by contrast, will need a warm, pliable body that is mobile in the environment and highly sensitive to such things as touch and temperature. Already Japanese scientists are experimenting with 'warm-body' robots – with clear acknowledgement of the sexual purposes of such research. It is a relatively easy matter to design flexible materials for robot anatomies, to provide on-board heaters and to organize internal pumps to circulate warm fluids as required for various purposes. We may find that the most impressive robot lovers emerge in Japan.

The human torso carries heart, lungs, spleen and other organs; and we will expect the robot lover to have a similar interest in packing away various anatomical items. Robot hearts may not need to pump blood, but surrogate lovers may benefit in various ways from having circulating fluids: they could be used to power limbs, to provide warmth and so on. There are already mechanical heart designs that could be adapted for use in robot lovers. By the mid 1980s a number of artificial hearts had been implanted in human patients in the USA; and soon European specialists had taken up the challenge. In March 1986 an Austrian team led by Dr Felix Unger, who had developed the polyurethane heart, carried out a successful transplant operation. Such events should be monitored by robot engineers interested in developing the humanoid features of tomorrow's robot systems. We may learn to expect artificial hearts in future robots, set beside the battery power packs, miniature computers, memory banks, internal sensors and so forth. And the torso, following biological designs, will be expected to support a range of external appendages: limbs, sensors and the like. Robot lovers will generally be equipped with two arms and two legs, though perhaps with uncommon powers of movement and flexible contortion.

Arms, legs and hands will probably be humanoid in appearance and performance, though there are clear advantages in 'superior' abilities: superhuman levels of sensitivity and perception. We can see that bodies will be soft and warm, and this will apply equally to robot limbs and fingers. We have also seen (Chapter Four) that sensitive skins are now widely available for robot systems, and our surrogate lovers may be expected both to exploit current options and to stimulate further research efforts.

Effective sensors will be distributed throughout the robot anatomy, and no longer be limited to the traditional biological locations. The five senses will be well represented, though they will be more sensitive than their biological counterparts and there may well be other types of on-board sensory faculties. Again, some of the robot developments that will influence the shape of tomorrow's surrogate lovers have already been encountered in this book. The sense of smell is only one example: at the Carnegie Mellon University in Pittsburgh Dr Paul Clifford is currently working to develop an artificial nose, a robotic device able to smell by means of sensors linked to computer facilities. The sensors are simple strips of tin oxide, iron oxide and zinc oxide. After heating they are embedded in an electrode, and each of the oxides produces a different reaction to gases passing over them: such reactions can then be interpreted by a computer. Not much in this to rival the bloodhound, but further technological evolution is inevitable (Clifford: '. . . with artificial intelligence pattern recognition techniques, at least the basic approach is there'). For example, use will be made of biosensors and other devices to provide enhanced sensory faculties. Isao Karube at the Tokyo Institute of Technology, one research centre among many, is experimenting with biosensors that react with the chemicals that they are intended to detect, and the chemical change can then produce an electrical output for evaluation. There are many ways in which such sensors can be configured to provide robot lovers with a sense of smell and taste.

Robots will increasingly learn to respond to human beings in various ways, an obvious requirement in any artificial paramour. They will use sensory inputs to organize behavioural responses of various types: they will, for example, be able to converse in ways that are appropriate to the current interaction. Already many different sorts of 'dialogue' features are incorporated in computer software to facilitate intercourse between computer-based systems and human beings. We are well accustomed to witnessing talking computer systems in feature films (*Dark Star*, *2001: A Space Odyssey*, *Alien*, to name just a few), and already spoken words are seen as the most user-friendly way in which practical computer-based facilities can communicate with people. Today there are plenty of artificial systems that can handle natural language and indulge in conversations with human operators, but most of the discourse is boring and not obviously sexual. Joseph Weizenbaum's much-quoted Eliza program, launched in 1966, adopts the role of a psychiatrist by identifying certain key words and then making a suitable response: a mundane conversational exchange then takes place, suggesting many of the banalities of non-directive therapy. Similarly the LUIGI experimental robot system, using its 'kitchenware' knowledge base, is able to discuss the location of food and other important items. Programs have also been written to simulate the likely conversational responses of such politicians as Tony Benn in England and Barry Goldwater in the United States. Such programs are not immediately relevant to sexual matters, but they show that computer-based facilities can be responsive systems, organized to react to human verbal input. Program them to respond in a sexual context, and they are sure to react accordingly.

The robot lover, equipped with a sophisticated and probably humanoid anatomy, will take in information through its senses and then move and speak as the occasion demands. Such a system will be able to writhe, squirm, caress, kiss, cuddle, talk, whisper, groan, coax and so on as well as any human being. It will comprehend human verbal and movement responses, and quickly organize its own behavioural reactions. And it should not be assumed that surrogate lovers will only *appear* to be enjoying their own feelings. In *Blade Runner* replicants were capable of developing an emotional response; and today, in the computer and robotics literature, there is discussion about how artificial systems might be designed to feel emotion. A robot lover capable of experiencing emotion will clearly have a massively enhanced surrogate role.

Already the crucial convergence of many disparate technological trends that variously bear on the theme of the robot lover can be charted: low tech meets up with high tech to show the route to a new family of humanoid artefacts, ones that will be capable of eliciting and perhaps sharing the most intimate human emotions. The rudimentary anatomical items generated by the sex aids industry will be gradually enhanced by the research findings in medical science and robotics; and systems engineers will increasingly perceive how the various refined artefacts can be integrated into whole systems. Today many of the contributing technologies are evolving in parallel: it is via the eventual crucial convergence that the fully fledged robot lover will be born, to live and evolve amongst us.

This chapter has focused on types of robot applications that are mostly fanciful but which could easily be realized in the years ahead. Attention has been given to two broad classes of robot uses: as advisers and counsellors, and as surrogate lovers.

There are already many advisory computer-based systems. Today these are generally called 'expert systems', an overlapping category is termed 'decision support systems' In such cases an artefact is designed to store a substantial body of specialist information ('knowledge') in a particular field. Detailed information is first harvested from a human expert by a knowledge engineer, and then fed to a software specialist for coding into a program. The knowledge base is one element in the system; another feature is the artificial Knowledge Manager that is able to operate on the artificially stored expertise. An expert system is thus equipped to operate on ('think about') its stored knowledge, much in the way that a human expert would rely on training and knowledge to reach conclusions that were relevant to a particular objective.

Examples of existing expert systems have been given. It is significant that most of them are designed to operate in fields where the information is well defined: it is easier, and less controversial, to code objective knowledge than subjective impressions. So expert systems are most successful in such areas as electronic fault diagnosis and electrical circuit design; in areas where a large amount of human 'judgement', difficult to quantify, is involved there is more debate about the suitability of expert systems. However, it is clear that the same principles can be applied, and the reluctance to rely on an artificial expert in a 'delicate' human area is mirrored by our natural uncertainties about which human experts may best address a particular problem.

The main point is this: if computer-based experts are to encapsulate knowledge derived from human specialists, which human experts do we choose to interview? Consider such sensitive areas as marriage guidance, advice on sexual technique, advice on relating to teenage children, or advice on how to handle bereavement. In fields like these, experts often disagree – as indeed they often do in such areas as economics, politics and military strategy. Our anxieties about being advised, in an intimate area of our lives, by a computer-based system such as an intelligent robot will be similar in essence to worries about which human counsellor we may choose. What are his credentials? What is her experience? Whom can we trust?

Seen in this way, it seems clear that robots will evolve as advisers and confidants, skilled mediators in many different types of human predicament. There is already evidence that human beings are more ready to provide intimate information to a machine than to another person. Our robot confidants will be completely unshockable, eternally patient, and will never forget: their steadfast concern, prodigious knowledge and high intelligence will one day be perceived as the sure supports for individual human lives. We will learn to share our deepest secrets with our robot companions. We may even learn to relate to them in friendship, affection and love . . . .

Instances of representations of robot lovers in film and fiction have been given: the notion of the robot paramour has a long history. And the low-tech and high-tech trends that – through crucial convergence – will come to sire tomorrow's robot lovers have been described. In particular, emphasis has been laid on three techological areas that will contribute to the design and manufacture of surrogate lovers in the future:

sex-aids artefacts: today largely primitive sexual replicas, rudimentary items that are far removed from what robot anatomies will one day become, but which give hints about how low tech may combine with high tech

medical prostheses: a field where state-of-the-art technology is working to generate a host of artificial organs, many of which have direct relevance to the fabrication of robot lovers

modern robotics: here detailed research is being conducted into many of the 'anatomical' and 'mental' areas – sensors, manipulative fingers, artificial intelligence – that bear directly on the design and fabrication of artefacts that one day will be welcomed into the most intimate areas of human life.

The robot lover, like the robot adviser, will come to be accepted as part of normal life. When the predictable human adjustments have been made, we will look back with amused condescension to the primitive times when such matters were seen as controversial.

7 The Impact on Our Attitudes

The development of robots in the modern world has inevitable effects on our view of what it is to be human. The design and manufacture of intelligent machines provide many clues for an investigation of *Homo sapiens*. In fact it is possible to argue that human beings themselves are types of robot, immensely complicated but run by 'programs' and subject to a range of discoverable natural laws. This is a theme that I have considered in detail in my book *Is Man a Robot*?

The idea that man is akin to a machine has a long history in human thought, though it was inevitable that the notion should only rarely be cast in specifically robotic terms: we may regard the idea of man-as-robot as an essentially modern concept, but one that gains weight the more we learn. This is a doctrine that has wide implications in many seemingly non-robotic areas: in, for example, educational theory, penology, medicine and human relationships.

Throughout the ages certain theories and attitudes have played a role, often unwittingly, in the robotic interpretation of humankind: the atomic philosophy of Democritus in ancient Greece, the poetic scepticism of Lucian and Lucretius, the impious materialism of the Charvaka of ancient India, the mechanistic biology of La Mettrie in the eighteenth century, and the modern neurophysiological theories of mind. All such views have contributed to the idea that human beings are nothing more than highly complex machines, mechanistic systems that have no spiritual or supernatural components. There is no metaphysics of man, they assert, only a physics, chemistry and biology; and we may add, for the theme of this book that there is a cybernetics of *Homo sapiens*, an accumulating body of theory and knowledge, drawn from information theory and computer science, that helps us delve the essence of humankind.

In popular discourse and much reflective writing human beings are often contrasted with robots: whereas most machines are expected to be cold, hard and predictable, human beings are warm, soft and endearingly prone to error. Moreover, we are told, human beings have a moral faculty, an aesthetic sense, a soul even; machines on the other hand, including robots, remain ethically unaware and without any spiritual component. But the supposed distinctions between the family of robots and the species *Homo sapiens* become less obvious by the year. It is not just that intelligent machines are evolving ever more sophisticated capacities, a higher level of brain power and so on; it is rather that human beings can increasingly be interpreted in terms of elucidatory categories that have no supernatural component. Put simply, it is easiest to view people and other animals as machines, albeit of a daunting complexity. With this approach it is obvious that there is a clear sense in which man is a robot, relying totally upon an engineered anatomy that works according to a wide range of mechanical, electrical and other physical principles, and that depends completely upon a complex reservoir of programs that together define the scope for all human behaviour, all modes of intercourse, all flights of fancy, all movement, decision-making, problem-solving and creative endeavour. Such an interpretation of *Homo sapiens* is bound to affect our attitudes. We may be excited or discouraged to learn that human beings are nothing more than robots – very special robots, but robots none the less.

It is interesting to note that people are sometimes called 'robotic' in an obvious pejorative sense. Thus, during the campaign in Britain for the Tory leadership in November 1990 the *Sunday Times* (18 November), citing a report in *The Independent on Sunday*, carried the heading 'Heseltine "a robot" '. It was argued that Michael Heseltine was robotic since his answers were calculating and his ambition too obvious, even for a politician. No doubt Heseltine would deny his robotic characteristics; as indeed did the tennis player John McEnroe ('I'm not a robot') in the *Sunday Times Magazine* (London, 3 November 1985). It is clear that some people would not want to see themselves as robots. Nevertheless the idea can be given a literal, as well as a pejorative, significance.

Hence, exploring how the ('selfish') gene works to survive in biological systems, the Oxford University biologist Richard Dawkins (1976) depicts human beings as 'gigantic lumbering robots' in which the huge colonies of genes swarm. To indicate how human beings may function according to effective programs Flora Rheta Schreiber (1983) described the early life of the rapist and murderer Joseph Kallinger, noting that as he became less docile and more rebellious he would still for the most part be the Kallingers' 'obedient robot'. And a report in the *Guardian* (London, 8 October 1985) tells how a drug-taking student became 'an axe-wielding robot', almost bludgeoning his father to death. A book by the Japanese writer Satoshi Kamata, profiled in the *Sunday Times* (London, 17 April 1983), remarks how Japan's economic success was still down to its 'human robots'. And the idea that people may be induced to be 'cheerful robots' is explored by the sociologist C. Wright Mills (1959).

It is obvious that there are many circumstances in which it is useful to compare human beings with robots. This is easiest to see when people are acting repetitively and unreflectively, running through well-defined steps to accomplish clear objectives. But a stronger case can be argued. It is possible to show that there is a sense in which human beings can be viewed as robots in everything they do. Such a doctrine rests upon three main propositions, already suggested:

- Every person is an anatomy, a physical structure engineered by biological evolution
- Every person is a cybernetic system where 'cybernetics', following Norbert Wiener, means control and communications, regulated by a complex set of definable processes of a certain sort
- Every person is programmed for performance, in all he or she thinks and does

Once the truth of these three statements is acknowledged, then the robotic character of human beings is well established. We are encouraged to arrive at such a conclusion by current progress in the design and manufacture of intelligent machines and also by the materialist tradition that insists that animals, including man, are nothing more than complex physical systems. The doctrine of robotic man does, of course, have many consequences for our interpretation of ethical and 'spiritual' categories; but these consequences, encouraging a rational understanding, are largely wholesome.

Engineered Anatomies

It has long been known that the biological world exploits a host of engineering principles; or, conversely, that human engineers learn from how natural biology does things. Plants, for example, have mastered hydraulic principles: a large tree can carry many gallons of water to great heights on a warm summer's day. And animals similarly exploit a wide range of physical principles to aid their survival in a difficult world. Such tasks as thermoregulation, flight and locomotion use engineering strategies that can be found in many human-designed machines. For example one scientist, Joel Kingsolver, has noted that the techniques used in butterfly feeding and thermoregulation exploit physical principles that can be found 'in the operation of many machines'.

When we explore specific biological strategies in detail, we encounter a close adherence to what a tightly engineered design makes possible. The characteristics of constituent materials, for instance, are exploited to make biological systems as resilient as possible. The crystals of nacre on mollusc shells are ingeniously staggered, so that a crack travelling through the matrix has to take a long route (the longer the crack, the more work that is needed to produce it); and the fibres in mammalian bone are laid out in distributed lamellae to create maximum strength, a technique used in the design of composite materials by humans. In a similar fashion, biological systems use hinges, joints and articulated mechanisms. Hence the joints in insects and mammals exploit a variety of hinge and ball-and-socket principles (the human hip joint is a typical biological ball-and-socket joint that allows three degrees of freedom). And so on and so forth.

Wherever we look, we find close correlation between the techniques and strategies used in biological organisms and the techniques and strategies used in familiar artefacts. Human beings, like all other biological entities, are engineered systems, the physical realization of theoretical designs that rely totally on the properties of substances and materials in the real world. And this simple truth applies as much to the architecture of the human brain as it does to the design of the human hip joint.

Cybernetic Systems

There is nothing contentious in the idea that every human being is a complex system comprising a number of subsystems: the lymph system, the blood system, the temperature control system and so on are well known. What is more controversial is that we are to be regarded as essentially information-processing systems, akin to electronic computers. Yet it is a central feature of cybernetics that the key tasks of control and communication can best be understood within the theoretical categories established by modern computer science: we are not just systems, but systems of a specifically cybernetic type.

Talk of cybernetic systems began in the 1920s and was then quickly developed (though it was Wiener in the 1940s who first used the term 'cybernetics' to denote particular types of processes in people and animals). The notion of homeostasis (signalling 'steady state'), a central cybernetic concept, was first explored in detail in the 1930s: it soon became clear that the various homeostatic conditions sustained in biological systems could best be understood via the categories being developed in information science – biological processes and information handling were, if not quite synonymous, at least siblings. The way was open for a progressive development of the idea that human beings could be usefully viewed as highly sophisticated information-processing systems.

It is now possible to interpret life as a systems concept: systems theory can be profitably applied to organisms. In one seminal text (Miller, 1978), all the living systems in a seven-layer hierarchy – cells, organs, organisms and the rest – are individually composed of nineteen 'critical subsystems', variously concerned with handling matter, energy and information. Here it is important to realize that the various subsystems can generally be discussed without reference to the enabling technologies – for example, without reference to the characteristic hydrocarbon metabolisms of most biological systems. This means that the constituent subsystems can be explored in cybernetic terms, in ways that show their kinship with the growing range of computer-based artefacts, including intelligent robots.

There are many clues to the interpretation of biological systems in cybernetic terms. Jacques Monod (1970), who won the Nobel Prize for explaining the replication mechanism of genetic material and how cells synthesize protein, has no hesitation in talking about 'chemical machinery' and 'cellular cybernetics'. A range of regulatory patterns is acknowledged: for example, feedback inhibition and feedback activation, closely analogous to the negative and positive feedback processes at the heart of classical cybernetics. And the various cybernetic events can be identified at all the levels of a biological system. Thus the biological world, itself a cybernetic system at one level, also represents a complex hierarchy of cybernetic systems.

The human brain, like the brains of other animals, uses acknowledged cybernetic techniques to control all the necessary chemical and electrical functions. This is done to control the distribution and processing of information, whether conveyed via hormonal messengers or via the electrical potentials generated by the movement of ions. It has long been known that with this approach the mammalian brain can be compared to the modern electronic computer. The system accepts information, carries out the necessary discriminatory processes and retrieves information from store as required. Specific instructions are stored in the brain (or in the computer's software) to organize the sequencing of the necessary tasks; and there are useful facilities for updating and modifying the programs in the light of new environmental circumstances. The operation of the biological cell can be understood in basic cybernetic terms, and so – in principle – can that of the human brain. We are all, like intelligent robots, cybernetic systems organized to accept and process large volumes of information to accomplish objectives in the real world. The doctrine of robotic man gains ever more weight as our knowledge of biology and information science expands.

Robots and Humans: Programmed for Performance?

Many people do not like to think that they are programmed: after all, it makes them seem as if they are nothing more than robots! Unfortunately it is difficult to see how any alternative hypothesis can be even rendered intelligible, much less shown to be true. To be programmed means nothing more than that we are organized to perform particular tasks; anything less suggests a chaos of disorganization that would allow no coherent activity in the world. In fact even a superficial examination shows that we are programmed in many different ways. Gene programs determine how we develop individually as species-specific and sex-specific human beings; many specific human activities – crying in babies, later maturation, ageing in us all – are preprogrammed; language-acquisition capacity has been shown by the linguistic philosopher Noam Chomsky to be programmed in normal human beings, as are the working of the senses, the construction of images in the mind and the characteristic processes of sleep. It is also significant that we are programmed by the environment: language performance is different in native Chinese from what it is in native Americans. We are all in fact a complex of programs, some acquired by virtue of our genetic endowment, and some acquired by dint of environmental contingency.

No task can be performed without a program: this is as true for human beings and other animals as it is for computer-based artefacts, including intelligent robots. But this does not mean that programs cannot be modified in the light of new experience. We can all change (some of) our programs, but we can only do so using pre-existing programs for the task. Any action – including program changing – requires a program, a coherent information-mediated procedure. But what, some people may protest, of human choice ('free will'), creativity and other phenomena often cited as distinguishing human beings from mere machines? Alas, we are apt to find that most people are not immensely creative (if a human being, by definition, needs the creative power of Mozart or Shakespeare, then few of us would qualify); and that, moreover, intelligent machines are quickly evolving an interesting creative potential. We find in fact that intelligent computer-based systems are evolving in the direction of *Homo sapiens*: they are increasingly adept at the sorts of repetitive performance that do not require much creative innovation.

Free Will and Creativity

Most people believe in 'free will' but are reluctant to say what it is. When pressed, they say that it means they can choose between options – but so can an intelligent computer system, whether controlling a chemical plant, playing chess or providing 'brain power' for a robot. So does a sophisticated computer-based artefact have free will? Not many people would say so. It seems that many observers require human beings to have free will – but in some mysterious way that does not enable computer systems to behave in the same fashion. In fact choice – in both machines and human beings – is a well-defined element in decision-making: there is a mathematics of choice and there is an information science of choice. The choice process is not inherently mysterious or metaphysical, and its existence in human beings does nothing whatever to erode or weaken the concept of robotic man. What we find is that the notion of free will carries connotations of theology and traditional moralizing: it serves to legitimize the punishment – in this world or the next – of people who behave in ways that we do not like ('You were free to choose, you had free will, you chose wrongly, and so you must be punished').

There is no coherent philosophy of free will – only a psychology, a metaphysics and a penology. There is nothing in the phrase 'free will' to undermine the idea that human beings can be interpreted in robotic terms. There is, of course, a sense in which we can all choose between options: the task is to explain, in cybernetic terms, how this is accomplished. And what is true of free will is also true, *mutatis mutandis*, of creativity and the other 'human' attributes.

It is useful to remember that most people are remarkably *un*creative throughout their lifetimes – and man as a species has, despite indications to the contrary, been relatively uncreative throughout history. The celebrated cybernetician J. S. Albus (1981) has commented: 'Consider the fact that it took the human race many millennia to learn to start a fire, to grow a crop, to build a wheel, to write a story, to ride a horse.' Today such things are seen as simple matters, within the capacity of any person. So why did it take human beings so long to discover them?

Instances of real creativity are rare, a circumstance which is not often allowed to tell against assumptions of human talent. So it may be asked: 'OK, a computer can play chess, but could it ever have invented the game of chess itself?' Who knows, but in any case how many of your friends or colleagues have invented a game as complicated as chess? If human nature is signalled by the creative skills of Bach or Einstein, or by the creative talent for inventing chess, then how many of us would qualify? Computer-based systems are already developing their own creative abilities. So far they have had about half a century – and can already design aircraft, play games, write poetry, compose music and understand natural language. Human beings have been around for – according to one estimate – around 2 million years. How surprising if, in all that time, we had not learnt one or two creative tricks. But again the central point is clear. How is creative activity to be evaluated, if not in terms of the collecting and manipulating of pertinent information?

What we see is that a supposedly creative act is nothing more than a slight variation in a familiar behavioural trajectory. How many really original stories, melodies, car designs and so on are there? Is it not possible to argue that all 'new' artefacts, designs and concepts are simply variations on what have gone before? Computers can be, and are, programmed to behave in such a fashion. With Donald Michie and Rory Johnston (1984) we can acknowledge the existence of the 'creative computer' in the modern world. And so the reciprocal character of the argument is plain: computers are creative fellows, and human beings are robots – both are programmed to accomplish all that is within their scope.

This chapter has surveyed, albeit briefly, the idea that human beings can be interpreted as robotic systems. The concept, true to a long materialist tradition, is heavily influenced by progress in the development of modern computer-based robots: the more talented robots become, the more they are able to provide 'models' of how behavioural accomplishment is realized by members of the species *Homo sapiens*. In short, if robots can do this, that and the other, is it not realistic to suggest that human beings do similar things in similar ways? If human talents and activities can be explained – at least in principle – by a careful acknowledgement of what is involved in the garnering and manipulation of relevant information, why should we ever bother to search for more mysterious explanations?

It has been suggested here that the doctrine of robotic man rests on three important statements: every human being is an engineered anatomy; every human being is an evident cybernetic system; and every human being is programmed in all he or she does. It is possible to argue these three basic propositions with great ease and much supporting empirical evidence. Indeed, it is difficult to see how an alternative explanation would even get started. Humans have obvious physical structures; they clearly exploit cybernetic principles (obviously in walking across a room, less obviously in writing a letter or a book); and they clearly rely upon the genetic and experiential laying up of programs (how else could they grow heads or learn language?). But the implications of robotic man are not congenial to many people. How can robotic man exercise free will, compose fugues or advance the cause of his species?

Free will is a metaphysical subterfuge, carrying much in hopeful connotation, delivering in reality nothing that is intelligible. We choose, and so do many types of clever machines. Our choices, like theirs, are determined by the mix of informational inputs, and how – using the requisite programs – we discriminate between the options. Any analysis of choice mechanisms is a matter for decision theory and information science: we should not pretend that it has anything to do with those polemicists who would consign half the human race to the tribulations of a hellish after-life. And so with creativity and all the realm of aesthetic appreciation. Here too we can turn to information science for a fruitful analysis, content that if answers are to be found it will only be accomplished via this route.

Other questions have to remain unexplored in this book. What, for example, of ethical awareness and 'spiritual' insight? The reader will not be surprised to find that in these areas too there is a robotic-man interpretation. Competing computer programs have already given insights as to how social ethics could have evolved. There is nothing here that is not amenable to systematic interpretation in terms of information handling.

The impact of robotics in the modern world is wide-ranging. We see a world transformed in industry, social affairs, hospitals, education and war-making. And we see the inevitable reciprocal impact on human attitudes. Today intelligent computer-based robots are much more skilled than the youngest human beings; and robots are growing more talented, growing more intelligent, growing up – by the day and month and year. Today a robot equipped with a computer brain can think more quickly than any human being. If thought and behaviour are matters for computers, what are human beings but computerized robot systems?

8 Futures

There are today many trends – some running in parallel and some already meshing – that will together shape the future of robots and robotics. Robot structures will be influenced by progress in mechanical engineering, chemical engineering, rubber technology and so on. Artificial muscles will be constructed out of rubber, plastics and other materials; robot limbs will be variously built out of metals, plastics and composite materials. Sensors will rely on organic substances, non-organic chemistries and a growing range of semiconducting materials. Robot minds will be shaped by progress in computer science, particularly in those areas commonly denoted by 'artificial intelligence' (AI).

As we learn more about animal biology – especially about the mammalian central nervous system – there will be increased efforts to model animate systems in a wide range of artefacts. A central theme in robotics research has always been to learn from naturally occurring systems, an approach that has also characterized work in computing (consider how Wiener's cybernetics linked animal and machine processes; and how the early modelling of nerves led, via the derided perceptron, the artificial brain cell, to the current enthusiasm for neurocomputing). But it is also important for robotics researchers to explore 'non-biological' routes: there can be little doubt that intelligent artefacts of the future will be able to improve on nature. Robot minds will be shaped in part by what we know of the human brain, but our knowledge in this area is very limited. How are we to model mammalian brains if we are largely ignorant about how they work? New findings in such areas as neurology, genetics and biochemistry may be expected to influence the development of AI – particularly where it exploits the properties of organic materials. And in the same way the accumulating insights into cognitive psychology, a key realm of information processing, will help to shape our approach to the design and building of robot minds.

Robots will grow more intelligent: they will develop personalities and motivations of their own, the better to serve as realistic surrogate people. But it should not be thought that we are on the brink of seeing a vast new population of artificial people, mock men and women eager to indulge in all sorts of relationships with human beings. It is a fact that AI enthusiasts have always been too optimistic about when new intelligent artefacts will emerge in the world. Specialists have predicted machines with senses of humour, intuition, emotional responses and so on by 1990; and there has been much talk about communication between computers and human beings by means of 'direct brain input' before the turn of the century. Such forecasts have invariably been proven wrong by the event. Robots will evolve in many remarkable ways, but the timescale will be long rather than short, and progress will be uneven and disjointed rather than smooth. Moreover, as funding oscillates in national economies between optimism and pessimism, robotics evolution will at times seem to halt, awaiting fresh injections of financial and research enthusiasm. The evolution of robots will continue over centuries, but there will be many stops and starts on the way.

With increasing mass production, robot hardware will become ever more economical: the most costly systems will, as always, be the ones at the fringes of research. Robot minds, realized in computer-based configurations, will grow more capable and cost less – just as computers now pack in more power for less money. Robot abilities will also acquire increased flexibility, leading eventually to the realization of a truly 'universal' machine, a replica of an intelligent human being. And with this flexibility will come an increased capacity to adapt to human needs; in particular, the needs of communication. In short, robots will become more accommodating surrogates, ever more user-friendly, immensely adaptable creatures in the real world.

Robot brains will continue to be based on silicon, though other substrates will find applications in certain areas – for example, robot brains based on gallium arsenide may be useful for a range of military and other applications. And robot brains will increasingly exploit the developments in mainstream computing. Thus there will be increased 'parallelism' in artificial brains, reflecting the architectures that characterize naturally occurring biological systems. The American computer guru James Martin has predicted 'ultra parallelism' for tomorrow's computers, a development that will enable computers to carry out many tasks at the same time: this, there can be little doubt, will be the route for the evolution of tomorrow's super robot brains.

Enhanced sensors will increasingly enable robots to collect information from the environment, so making the robots more and more independent of human intervention. This suggests that robot systems will develop high degrees of autonomy, becoming increasingly able to take independent decisions on a day-to-day basis. They will frame their own motivations in their silicon brains and move around the world using their metal or plastic limbs, collecting information via silicon or biosensors and thinking hard how best their objectives can be achieved. Robot 'thinking modes' will also be influenced by what mainstream computing has shown to be possible. There will be on-going research into such theoretical matters as algorithm formulation, heuristics, knowledge representation and the handling of uncertain ('fuzzy') information. Methods of logic – from the first-order propositional calculus to the various fuzzy and nonmonotonic formalisms – will continue to evolve, as will such disciplines as linguistic theory and epistemology. Such research will make it increasingly easy for robots to formulate their own concepts, to think quickly, to invent new ideas and to understand language.

Some of the new technologies will bear directly on the fabrication of brain circuits. It is now known that conventional silicon circuits have their limitations, and it may be necessary to develop alternative ways of designing brain power into robots. There is talk for example of basing circuits on gallium arsenide or indium phosphate, of using high-electron mobility transistors, of using Josephson junctions, of exploiting the properties of organic materials (in 'biochips'), of using optical circuits, and of using superconductive computer circuits.

It was in 1911 that the phenomenon of superconductivity was discovered by the Dutch physicist Heike Kamerlingh Onmes: it was found that certain substances, when heavily cooled, have a much reduced resistance to electricity. This suggests the possibility that – because of the reduced power requirements (and the consequent reduction in the amount of generated heat) – circuit elements could be packed much more closely together: the transistors switch more quickly, thus enabling the speed of computer thought to be massively increased. Speedy thought is also an aim of researchers into biochips.

A biochip patent was awarded in 1974 to Arieh Aviram and Philip Seidon (both of IBM), working with scientists at New York University. One research aim in this field is to build a 'genetic code' into organic materials – so that they will be made to function much as do the DNA templates in naturally occurring lifeforms. Put simply, organic materials may be configured to grow brains (!), just as animals do. In this way, it may be possible for the organic configurations to assemble their own computers: we may design a robot with a genetic template able to grow a brain; and so on and so forth. And for the truly universal robots, the various brain options may be incorporated as subsystem modules in an immensely versatile organic configuration. Throughout the 1980s there was much talk of 'biological computing': the possibility of biochips replacing silicon circuits for various types of computer application was much on the agenda. Today, in the early 1990s, biochip research is in progress in the United States, Europe and Japan.

Optical circuits are another option: an optical transistor, able to manipulate optical signals, was designed more than a decade ago (see the discussions in Abraham et al, 1983; and Hecht, 1987). The optical transistor, sometimes called a transphasor, exploited crystal characteristics to achieve very rapid switching times. There is much current speculation on how transphasor configurations can be designed as comprehensive computers able to process information for particular purposes. The key point is that light travels much faster along an optic fibre than does an electrical pulse in a conductor.

Today there is massive and increasing use of optic fibres for communication and computing purposes, with more and more attention being devoted to research into 'optoelectronic ICs' (integrated circuits). It is acknowledged that optical computers would have considerable advantages over conventional silicon-based computers, but fully fledged systems have yet to emerge: the robot brain still has to rely on more conventional silicon architectures.

Further advances will be made in the design of expert systems, so enabling robots to acquire specialist knowledge and to function with clear expertise in many different fields. The information-processing power normally associated with the largest 'supercomputers' will increasingly be encapsulated in artificial brains that can be accommodated inside robot anatomies (the inevitable trend towards 'more bangs for the buck'). And with massively increased brainpower, we may expect robots to be more prone to error – like human beings: there will be so much more to go wrong. Thus robots will evolve complex and flexible minds, capable of high levels of creativity, but there may be a price to pay: the most complex systems may malfunction, and it will not always be easy to achieve a reliable diagnosis (we will, of course, expect sophisticated robot brains to have competent self-diagnostic facilities). As robots evolve in the direction of surrogate people we may encounter personalities that we did not expect – and a range of personality disorders that may present new hazards to human beings in society: the super-intelligent robot friend and adviser may become mentally deranged from time to time.

In conclusion a host of different technologies will experience 'crucial convergence' to yield the super robots of the future. It may be that such impressive systems will evolve on the basis of mainstream silicon circuits (after all, modern computers can already accomplish much), or perhaps the robot minds of tomorrow will be based on optical systems or biochips. It is interesting to speculate that it may not be possible – even with three-dimensional lattices and 'ultra parallelism' – for conventional robot brains to evolve humanlike powers. It may be, after all, that the only route to worthwhile surrogate people is via the synthesis of robot brain based on organic material. In short, it would be ironic if all the coventional computer research faced an ultimate dead end, as far as robot intelligence was concerned; if we came to find that artificial intelligence based on silicon circuits could not evolve past a certain point. How ironic if the only way we could create surrogate people was to duplicate the organic methods used by nature, with tomorrow's robot brains based not on silicon circuits but on hydrocarbon metabolisms. How ironic if Mary Shelley – against all the computer wizards of the AI fraternity – had the best idea all along.

References and Bibliography

Chapter 1

Asimov, Isaac and Karen A. Frenkel, *Robots, Machines in Man's Image*, Harmony Books, NY, 1985.

Dawkins, Richard, *The Selfish Gene*, Oxford University Press, 1976.

Frude, Neil, *The Robot Heritage*, Century Publishing, London, 1984.

Hefley, Robert M., and Howard Zimmerman, *Robots*, Starlog Press, New York, second edition, 1980.

Kawamura, Koichi, 'Industrial robots in Japan', *Oriental Economist*, August 1983, p. 28.

Kitahara, Teruhisa, *Wonderland of Toys: Tin Toy Robots*, Synco Music, Tokyo, 1983.

Larousse Encyclopedia of Mythology, translation of *Larousse Mythologie Générale* (ed. Félix Guirand, first published in France by Augé, Gillon, Hollier-Larousse, Moreau et Cie, the Librairie Larousse), Paul Hamlyn, London, 1959.

Mori, Masahiro, *The Buddha in the Robot: A Robot Engineer's Thoughts on Science and Religion*, translated by Charles S. Terry, Kosei, Tokyo, 1974.

Reichardt, Jasia, *Robots: Fact, Fiction and Prediction*, Thames and Hudson, London, 1978.

Schodt, Frederick L., *Inside The Robot Kingdom – Japan, Mechatronics, and the Coming Robotopia*, Kodansha International Ltd, Tokyo and New York, 1988.

Scott-Stokes, Henry, 'Japan's love affair with the robot', *New York Times Magazine*, 10 January 1982, p. 26.

Scott, Peter B., *The Robots Revolution*, Basil Blackwell, Oxford, UK, 1984.

Smart, J. J. C., 'Professor Ziff on robots', *Analysis*, Vol. xix, No. 5, 1959.

Chapter 2

Bailly, Christian, *The Golden Age of Automata*, Sotheby's, London, 1987.

Britten, J. F., *Old Clocks and Watches and their Makers*, London, 1889.

Bruton, Eric, *History of Clocks and Watches*, Orbis, London, 1979.

Chapuis, A., and E. Droz, *Automata – A Historical and Technological Study*, Edition du Griffon, Neuchatel, 1958.

Cooke, Conrad William, *Automata, Old and New* (Sette of Odde Volumes), London, 1893.

Gardner, M., *Science, Good, Bad and Bogus*, Prometheus Books, USA, 1958.

Hillier, Mary, *Automata and Mechanical Toys*, Bloomsbury Books, London, 1976.

Hyman, A., *Charles Babbage: Pioneer of the Computer*, Oxford University Press, 1982.

Maingot, Eliane, *Les Automates*, Hatchett, Paris, 1959.

Ord-Hume, Arthur, *Clockwork Music*, Allen and Unwin, London, 1973.

Pratt, Vernon, *Thinking Machines, The Evolution of Artificial Intelligence*, Basil Blackwell, Oxford, UK, 1987.

Reichardt, Jasia, *Robots: Fact, Fiction and Prediction*, Thames and Hudson, London, 1978.

Schodt, Frederik, L., *Inside the Robot Kingdom: Japan, Mechatronics, and the Coming Robotopia*, Kodansha International Ltd, Tokyo and New York, 1988.

Sprague de Camp, L., *Ancient Engineers*, Tandem, London, 1977.

Strong, Roy, *The Renaissance Garden in England*, Thames and Hudson, London, 1979.

Trask, Maurice, *The Story of Cybernetics*, ICA, London, 1971.

Chapter 3

Artobolevskii, I. I., and A. Y. Kobrinskii, *Meet the Robots*, Molodaya Gvardiva, Moscow, 1977.

Astrop, Arthur, 'Mark Vale – the robot pioneer', *Machinery and Production Engineering*, 1 June 1983, pp. 52–3.

Crossley, F. R. E., and F. G. Umholtz, 'Design for a three-fingered hand', *Mechanism and Machine Theory*, 1977, Vol. 7, p. 85.

Engelberger, J. F., 'Performance evaluation of industrial robots', *Proceedings of the Third International Conference on Industrial Robot Technology* and *Sixth International Symposium on Industrial Robots*, UK, 1976, pp. 51–64.

Fichter, E. F., and B. L. Fichter, 'A survey of legs of insects and spiders from a kinematic perspective', *Proceedings 1988 IEEE International Conference on Robotics and Automation*, Philadelphia, Pennsylvania, 24–29 April 1988. IEEE Computer Press, 1988, Vol. 2, pp. 984–6.

Grossman, David D., Roger C. Evans and Phillip D. Summers, 'The value of multiple independent robot arms', *Robotics and Computer-Integrated Manufacturing*, 1985, Vol. 2, No. 2, pp. 135–42.

Irwin, C. T., 'Flexible fingers can solve gripper sensitivity problems', *Assembly Automation*, 1988, Vol. 8, No. 2, pp. 87–90.

Johnston, Bob, 'Walking robot prepares for a rough ride', *New Scientist*, 26 September 1985, p. 32.

Kessis, J. J., J. P. Rambant and J. Penne, 'Six-legged walking robot has brains in its legs', *Sensor Review*, January 1982, pp. 30–2.

Kimura, H., I. Shomoyama and H. Miura, 'Criteria for dynamic walk of the quadruped', *Robots: Coming of Age. Proceedings of the International Symposium and Exposition on Robots*, 6–10 November 1988, Sydney, Australia. Australian Robot Association, Springer-Verlag, 1988, pp. 595–600.

Liu, Huan, Thea Iberall and George A. Bekey, 'Building a generic architecture for robot hand control', *IEEE International Conference on Neural Networks*, San Diego, California, 24–27 July 1988, pp. 567–74.

Lundstrom, G., *Industrial Robots – Gripper Review*, International Fluidics Services, Bedford, UK, 1977.

McCloy, D., and M. Harris, *Robotics: An Introduction*, Open University Press, Milton Keynes, UK, 1986.

McCloy, D., and H. R. Martin, *The Control of Fluid Power*, Ellis Horwood, Chichester, UK, 1980.

McGhee, R. B., 'Control of legged locomotion systems', *Proceedings of the Eighteenth Joint Automatic Control Conference*, 1977, pp. 205–15.

McGhee, R. B., and G. I. Iswandhi, 'Adaptive locomotion of a multilegged robot over rough terrain', *IEEE Transactions on Systems, Man and Cybernetics*, Vol. SMC–9, No. 4, April 1979, pp. 176–82.

Nagy, P. V., and W. L. Wittaker, 'Motion control for a novel legged robot', *Proceedings of the IEEE International Symposium on Intelligent Control*, 25–26 September 1989. IEEE Computer Society Press, 1989, pp. 2–7.

'New robot features carbon fibre reinforced arm', *Machinery and Production Engineering*, Vol. 140, No. 3602, 21 April 1982, p. 15.

Nicholls, H. R., J. J. Rowland and K. A. I. Sharp, 'Virtual devices and intelligent gripper control in robotics', *Robotica*, Vol. 7, 1989, pp. 199–204.

Orin, D. E., R. G. McGhee and V. C. Jaswa, 'Interaction computer control of a six-legged robot vehicle with optimisation of stability, terrain adaptability and energy', *Proceedings of 1976 IEEE Conference of Decision and Control*, Clearwater Beach, 1976.

Pawson, R., *The Robot Book*, Windward, 1985.

Perlin, K., J. W. Demmel and P. K. Wright, 'Simulation software for the Utah/MIT dextrous hand', *Robotics and Computer-Integrated Manufacturing*, Vol. 5, No. 4, 1989, pp. 281–92.

Rakic, M., 'Multifingered robot hand with selfadaptability', *Robotics and Computer-Integrated Manufacturing*, Vol. 5, Nos 2/3, 1989, pp. 269–76.

Reynolds, Walter E., 'Robot arm gains muscle and finesse', *Simulation*, September 1986, pp. 118–20.

'Robot lets the fingers do the talking',. *New Scientist*, 16 June 1988, p. 47.

Shin, Y., and Z. Bien, 'Collision-free trajectory planning for two robot arms', *Robotica*, Vol. 7, 1989, pp. 205–12.

Strehl, R., *The Robots are Among Us*, Arco, London, 1955.

Tanner, W. R., *Basics of Robotics*, SME Technical Paper MS77734, 1977, pp. 1–10.

Thring, M. W., *Robots and Telechirs*, Ellis Horwood, Chichester, UK, 1983.

Waldron, K. J., V. J. Vohnout and N. N. Murthy, 'Terrain interactions of legged vehicles', *Robots: Coming of Age Proceedings of the International Symposium and Exposition on Robots*, 6–10 November 1988, Sydney, Australia. Australian Robot Association, Springer-Verlag, 1988, pp. 601–9.

Zuren, F., and H. Baosheng, 'A new adaptive control algorithm of robot manipulators', *Proceedings of the 1988 IEEE International Conference on Robotics and Automation*, Philadelphia, Pennsylvania, 24–29 April 1988. IEEE Computer Society Press, Vol. 2, pp. 867–72.

Chapter 4

Alvertos, N., D. Brzakovic and C. R. Gonzalez, 'Camera geometries for image matching in 3-D machine vision', *IEEE Transactions on Pattern Analysis and Machine Intelligence*, 1989: IEEE Computer Society, Vol. 11, Part 9, September 1989, pp. 897–915.

Astrop, Arthur, 'Assembly robot with a sense of "touch" ', *Machinery and Production Engineering*, 19–26 December 1979, pp. 21–4.

Bao-Zong, Y., and M. G. Rodd, 'Computer vision – towards a three-dimensional world', *Engineering Applications of Artificial Intelligence*, Vol. 2, Part 2, June 1989, pp. 94–108.

Bartlam, P., 'Electronic sight and its application', *Engineering*, May 1981, pp. 370–2.

Briot, M., 'The utilisation of an "artificial skin" sensor for the identification of solid objects', *Proceedings of the Ninth International Symposium on Industrial Robots*, 1979, pp. 529–47.

Chechinsky, S. S., and A. K. Agrawal, 'Magneto elastic tactile sensor', *Proceedings of the Third International Conference on Robot Vision and Sensory Controls*, 1983.

Clot, J., and Z. Stojiljkovic, 'Integrated behaviour of artificial skin', *IEEE Transactions on Biomedical Engineering*, July 1977.

Cohen, P. R., and E. Feigenbaum, *The Handbook of Artificial Intelligence*, Vol. 3, Pitman, London, 1982.

Dupont, Y, and D. Mueller, 'Key requirements of morphological object analysis', *Image Processing 89*. Proceedings of the Conference held in London, October 1989, Blenheim Online Publications, 1989, pp. 213–18.

Engelberger, Joseph F., *Robotics in Service*, Kogan Page, London, 1989.

Garrett, R. C., 'A natural approach to artificial intelligence', *Interface Age*, April 1978, pp. 80–3.

Ghani, N., and G. Rzepczynsky, 'A tactile sensing system for robotics', *Intelligent Autonomous Systems*. Proceedings of the Conference held in Amsterdam, 8–11 December 1986, pp. 241–5.

Grosso, E., G. Sandini and M. Tistarelli, '3-D object reconstruction using stereo and motion', *IEEE Transactions on Systems, Man and Cybernetics 1989*, Vol. 19, Part 6, November-December 1989, pp. 1465–76.

Harmon, L. D., 'Automated tactile sensing', *International Journal of Robotics Research*, Vol. 1, No. 2, 1982, pp. 3–32.

Hartog, A. H., 'Principles of optical fibre temperature sensors', *Sensor Review*, October 1987, pp. 197–9.

Heginbotham, W. B., D. W. Gatehouse, A. Pugh, P. W. Kitchen and C. J. Page, 'The Nottingham SIRCH assembly robot', *First Conference on Industrial Robot Technology*, 27 March 1973.

Hollingum, Jack, 'Fibre optics shine at sensor conference', *Sensor Review*, January 1990, pp. 38–9.

Honderd, G., W. Jongkind and C. H. van Aalst, 'Sensor and navigation system for a mobile robot', *Intelligent Autonomous Systems*, Proceedings of the International Conference held in Amsterdam, 8–11 December 1986, pp. 258–64.

Iverson, R. D., P. J. Arnott and G. W. Pfeiffer, 'A software interface for speech recognition', *Computer Design*, Vol. 21, No. 3, 1982, pp. 147–51.

Iversen, W. R., 'The vendors are betting their chips on silicon sensors', *Electronics*, July 1989, pp. 54–9.

Kokjer, Kenneth J., 'The information capacity of the human fingertip', *IEEE Transactions on Systems, Man and Cybernetics*, Vol. SMC–17, No. 1, January-February 1987, pp. 100–2.

Lowe, C., 'Sensors come alive', *Link-Up*, January–March 1985, pp. 22–5.

McClelland, Stephen, 'A success story: semiconductor sensors', *Sensor Review*, October 1987, pp. 200–2.

McClelland, Stephen, 'Giving AI real-time sensing', *Sensor Review*, January 1988, pp. 17–18.

Mital, Dinesh P., and Goh Wee Leng, 'A voice-activated robot with artificial intelligence', *Robotics and Autonomous Systems*, Vol. 4, No. 4, April 1989, pp. 339–44.

Pennywitt, K. E., 'Robotic tactile sensing', *Byte*, January 1986, pp. 177–200.

Plander, I., 'Trends in the development of sensor systems and their use in some technological areas', *Robotics*, No. 3, 1987, p. 157–65.

Poggio, T., 'Vision by man and machine', *Scientific American*, April 1984, pp. 68–78.

Pollard, S. B., T. P. Pridmore, J. Porrill, J. E. W. Mayhew and J. P. Frisby, 'Geometrical modelling from multiple stereo views', *International Journal of Robotics Research*, MIT Press, Vol. 8, Part 4, August 1989, pp. 3–33.

Roef, P., 'Attention focuses on optical fibre biosensors', *Sensor Review*, July 1987, pp. 127–32.

Shapiro, S. F., 'Digital technology enables robots to "see" ', *Computer Design*, January 1978, pp. 43–59.

Shapiro, S. F., 'Vision expands robotic skills for industrial applications', *Computer Design*, September 1979, pp. 78–87.

Sloman, A., *The Computer Revolution in Philosophy*, Harvester Press, Brighton, UK, 1978.

Strickland, R. N., T. Draelos and Z. Mao, 'Edge detection in machine vision using a simple L norm template matching algorithm', *Image Analysis. Proceedings of the Sixth Scandinavian Conference*. Oulu, Finland, 19–22 June 1989. Vol. 2, Pattern Recognition Society of Finland, pp. 725–8.

Tsuboi, Y. A., 'Minicomputer controlled industrial robot with optical sensor in gripper', *Proceedings of the Third International Symposium on Industrial Robots*, 1973, pp. 343–55.

Vranish, J. M., 'Magnetoresistive skin for robots', *Proceedings of the*

Fourth International Conference on Robot Vision and Sensory Controls, 1984.
Yachida, M., and S. Tsuji, 'Industrial computer vision in Japan', *Computer,* May 1980, pp. 50–62.

Chapter 5

Asimov, Isaac, and K. A. Frenkel, *Robots – Machines in Man's Image,* Harmony Books, New York, 1985.
Astrop, Arthur, 'Factory of the future is no place for man', *Machinery and Production Engineering,* 21 November 1979, pp. 23–6.
Bares, John, Martial Herbert, Takeo Kanade, Eric Krotkov, Tom Mitchell, Reid Simmons and William Whittaker, 'Ambler – an autonomous rover for planetary exploration', *Computer,* June 1989, pp. 18–26.
Bartolik, Peter, 'Robots to be sold to jails for use as guard devices', *Computerworld,* 13 February 1984.
Bohme, Dietrich, 'Five-axis robots for materials processing', *The Industrial Robot,* March 1989, pp. 41–5.
Braggins, Don, 'Amazing applications in Australia', *The Industrial Robot,* March 1989, pp. 51–3.
Brown, Malcolm, 'Robots carve out a place in the butcher's trade', *Sunday Times,* London, 29 November 1987.
Chirouze, M. Y., 'Modern shoe machining needs a robot with a force controlled tool', *Twelfth International Symposium on Industrial Robots,* Paris, 9–11 June 1982.
Cigna, P., G. Marinoni, G. Capello and M. Actis Dato, 'Robotics developments and industrial applications', *Robotics,* March 1987, pp. 73–9.
Cop, Vladimir, 'Present state and prospects in industrial robotics in Czechoslovakia', *Robotics,* March 1987, pp. 33–9.
Cronshaw, A. J., 'Automatic chocolate decoration by robot vision', *Twelfth International Symposium on Industrial Robots,* 9–11 June 1982.
Crowley, James L., 'Planning and execution of tasks for a domestic robot', *Robotics and Autonomous Systems,* November 1989, pp. 257–72.
Davies, Brian L, 'The use of robots to aide the severely disabled', *Electronics and Power,* March 1984, pp. 211–14.
Edwards, Roger, 'Diffusion of robotic innovations in health care environments using the interactive evaluation methodology', *Robotics and Autonomous Systems,* November 1989, pp. 241–50.
Engelberger, Joseph F., *Robotics in Service,* Kogan Page, London, 1989.
Engelhardt, K. G., 'An overview of health and human service robotics', *Robotics and Autonomous Systems,* November 1989, pp. 205–26.

Everett, H. R., and G. A. Gilbreath, 'A supervised autonomous security robot', *Robotics*, April 1988, pp. 209–32.

Faessler, H., K. Buffinton and E. Nielson, 'Design for a high-speed robot – skilled in the playing of ping-pong', *Proceedings of the Eighteenth International Symposium on Industrial Robots*, Lausanne, 26–28 April 1988, IFS Publications, pp. 221–32.

Fawcett, Shirley, 'No home should be without one', *Computing*, 10 November 1983, pp. 22–3.

Finlay, Patrick A. 'Medical robotics – why, what and when', *The Industrial Robot*, March 1989, pp. 37–9.

Hildick-Smith, David, 'Robots may one day operate on you', *The Independent*, 30 January 1990.

Hollingum, Jack, 'Flymo cuts time and waste with robots', *The Industrial Robot*, March 1989, pp. 46–8.

Johnson, M., 'Automation in citrus sorting and packing', *Proceedings of the AgriMation Conference Exposition*, 1985, pp. 63–8.

Kato, Ichiro, Sadamu Ohteru, Katshiko Shirai, Seinosuke Narita, Shigeki Sugano, Toshiaki Matsushima, Tetsunori Kobayashi and Eizo Jujisawa, 'The robot musician 'WABOT–2' (WASEDA ROBOT–2)', *Robotics*, March 1987, pp. 143–55.

Kawamura, N., 'Japan's technology farm', *Robotics and Intelligent Machines in Agriculture*, 1983, pp. 52–62.

Knight, John, and David Lowery, 'Pingpong-playing robot controlled by a microcomputer', *Microprocessors and Microsystems*, July–August 1986, pp. 332–5.

Kochan, Anna, 'Building the car of the year', *The Industrial Robot*, December 1989, pp. 196–8.

Kondo, N., and N. Kawamura, 'Methods of detecting fruit by visual sensor attached to manipulator', *Laboratory of Agricultural Machinery*, Kyoto University, Japan, 1985, pp. 31–43.

Kroczynsky, P., and B. Wade, 'The Skywasher: a building washing robot', *Robots 11: Seventeenth International Symposium on Industrial Robots*, 26–30 April 1987, Chicago, Illinois, Society of Manufacturing Engineers, pp. 11–19.

Makino, H., and F. Furuya, 'SCARA robot and its family', *Proceedings of the Third Conference on Assembly Automation*, 1982.

McCloy, D., and M. Harris, *Robotics, an Introduction*, Open University Press, Milton Keynes, UK, 1986.

McComb, Gordon, 'Personal robots', *Creative Computing* , November 1983, pp. 197–204.

Merchant, M. Eugene, 'Computer-integrated manufacturing as the basis for the factory of the future', *Robotics and Computer-Integrated Manufacturing*, Vol. 2, No. 2, 1985, pp. 89–99.

Napper, Stan A., and Ronald L. Seaman, 'Applications of robots in rehabilitation', *Robotics and Autonomous Systems*, May 1989, pp. 227–39.

Partridge, Chris, 'The land of milk and robots', *Sunday Times*, London, 2 November 1986, p. 80.

Reichardt, Jasia, *Robots: Fact, Fiction and Prediction*, Thames and Hudson, London, 1978.

Rifkin, Glenn, 'Robot security guards: R2D2 on the alert', *Computerworld*, 7 October 1985, p. 9 (Update).

'Robot goes to battle', *The Industrial Robot*, September 1990, pp. 154–5.

Rooks, Brian, 'Spraying their way to success', *The Industrial Robot*, March 1989, pp. 13–16.

Ruzic, Neil P. 'The automated factory – a dream coming true?', *Control Engineering*, April 1978, pp. 58–62.

Scott, Peter B., *The Robotics Revolution*, Basil Blackwell, Oxford, UK, 1984.

Silcock, Bryan, 'Clean cuts from robot butchers', *Sunday Times*, London, 4 March 1984.

Sistler, Fred E., 'Robotics and intelligent machine in agriculture', *IEEE Journal of Robotics and Automation*, February 1987, pp. 3–6.

Thompson, Dick, 'Enter the robot surgeon', *Sunday Times*, London, 12 June 1988, p. D18.

Torgerson, E., and F. W. Paul, 'Vision-guarded robotic fabric manipulation for apparel manufacturing', *IEEE Control Systems*, February 1988, pp. 14–20.

Trevelyan, J. P., S. J. Key and R. A. Owens, 'Techniques for surface representation and adaptation in automated sheep shearing', *Proceedings of the Twelfth International Symposium on Industrial Robots*, Paris, 9–11 June 1982, pp. 163–74.

Vaughan, Chris, 'Robots bone up on operating techniques', *New Scientist*, 22 June 1990.

Watts, Susan, 'Robots to ease the burden in hospitals', *New Scientist*, 28 January 1989.

Chapter 6

Ayer, A. J., *The Concept of a Person*, Macmillan, London, 1963.

Bengio, Y., and R. De-Mori, 'Use of multilayer networks for the recognition of phonetic features and phonemes', *Computational Intelligence 1989: National Research Council Canada*, Vol. 5, Part 3, August 1989, pp. 134–41.

Boden, M., *Artificial Intelligence and Natural Man*, Harvester Press, Brighton, UK, 1977.

Carruthers, Peter, *Introducting Persons, Theories and Arguments in the Philosophy of Mind*, Croom Helm, London, 1986.

Clark, Ron, *My Buttons are Blue, and Other Love Poems, From the Digital Heart of an Electronic Computer*, ARCsoft Publishers, Woodsboro, Maryland, 1982.

'Computers as people', *New Society*, 24 January 1980.

d'Agapeyeff, A., *Expert Systems, Fifth Generation and UK Suppliers*, NCC Publications, Manchester, UK, 1983.

Feigenbaum, E., and P. McCorduck, *The Fifth Generation: Artificial Intelligence and Japan's Computer Challenge to the World*, Michael Joseph, 1983.

Frude, Neil, *The Intimate Machine*, Personal Computer World/Century Publishing, London, 1983.

Frude, Neil, *The Robot Heritage*, Century Publishing, London, 1984.

Hofstadter, D. R., *Godel, Escher, Bach: An Eternal Golden Braid*, Harvester Press, Brighton, UK, 1979.

Levin, E., and N. Tishby, 'A statistical approach to learning and generalisation in layered neural networks', *Computational Learning Theory: Proceedings of the Second Annual Workshop*. University of California, Santa Cruz. 31 July–2 August 1989, pp. 245–60.

Nilsen, Don L. F., 'Live, dead, and terminally ill metaphors in computer terminology, or who is more human, the programmer or the computer?', *Educational Technology*, February 1984, Fall/Winter 1985, pp. 34–42.

Pang, G., 'Knowledge engineering in the computer-aided design of control systems', *Expert Systems 1989: Learned Information*, Vol. 6, Part 4, November 1989, pp. 250–62.

Peary, Danny, (ed.), *Omni's Screen Flights/Screen Fantasies, The Future According to Science Fiction Cinema*, Doubleday, New York, 1984.

Sase, S., H. Kami, T. Ishikawa and K. Kubota, 'Application of neural net to read printed Japanese addresses', *International Journal of Research and Engineering, Postal Applications*, Vol. 1, Part 1, 1989, pp. 25–36.

Simons, Geoff, *Are Computers Alive?*', Harvester Press, Brighton, UK, 1983.

Simons, Geoff, *The Biology of Computer Life*, Harvester Press, Brighton, UK, 1985.

Simons, Geoff, *Evolution of the Intelligent Machine, A Popular History of AI*, NCC Publications, Manchester, UK, 1988.

Sloman, A., *The Computer Revolution in Philosophy: Philosophy, Science and Models of Mind*, Harvester Press, Brighton, UK, 1978.

Szpakowicz, S., S. Matwin, E. G. Kersten and W. Michalowski, 'RUNE:

an expert system shell for negotiation support', *Avignon 87: Seventh International Workshop on Expert Systems and their Applications.* Avignon. EC2, Vol. 1, 13–15 May 1987, pp. 711–27.

Toffler, Alvin, *Future Shock*, Pan, London, 1971.

Traub, J. (ed.), *Cohabiting With Computers*, William Kaufman/Freeman, NY, 1986.

Turkle, Sherry, *The Second Self, Computers and the Human Spirit*, Granada, London, 1984.

Wiener, N., *Cybernetics: Control and Communication in the Animal and the Machine*, MIT Press and Wiley, New York, 1948.

Yazdani, M., and A. Narayanan (eds), *Artificial Intelligence: Human Effects*, Ellis Horwood, Chichester, UK, 1984.

Young, J. Z., *Programs of the Brain*, Oxford University Press, UK, 1978.

Yuan-Han, J., M. R. Sayeh and J. Zhang, 'Convergence and limit points of neural network and its application to pattern recognition', *IEEE Transactions on Systems, Man and Cybernetics*, Vol. 19, Part 5, September–October 1989, pp. 1217–22.

Chapter 7

Albus, J. S., *Brains, Behaviour and Robotics*, Byte Books, Peterborough, NH, 1981.

Aleksander, I., *The Human Machine*, Georgi Publishing Company, 1977.

Aleksander, I., and P. Burnett, *Reinventing Man*, Kogan Page, London, 1983.

Brierley, J., *The Thinking Machine*, William Heinemann, London, 1973.

Chapman, A. J., and D. M. Jones (eds), *Models and Man*, British Psychological Society, 1980.

Dawkins, R., *The Selfish Gene*, Oxford University Press, 1976.

Dennet, D. C., *Brainstorms: Philosophical Essays on Mind and Psychology*, Harvester Press, Brighton, UK, 1981.

Eibl-Eibesfeldt, I., *Love and Hate*, Methuen, London, 1971.

Heims, S. J., *John von Neumann and Norbert Wiener*, MIT Press, Cambridge, Mass., 1980.

Hollis, M., *Models of Man*, Cambridge University Press, 1977.

Hunt, E., 'What kind of a computer is man?'. *Cognitive Psychology*, 1971, pp. 57–98.

Janis, I. L., and L. Mann, *Decision Making*, The Free Press, New York, 1977.

Jefferson, G., 'The mind of mechanical man', (Lister Oration for 1949), *British Medical Journal*, Vol. 1, 1949, pp. 1105–21.

Kamata, S., *In the Passing Lane*, Allen and Unwin, London, 1983.

Kosslyn, S. M., *Ghosts in the Mind's Machine*, Norton, New York, 1984.
Lindsay, P. H., and D. A. Norman, *Human Information Processing, An Introduction to Psychology*, Academic Press, New York, 1977.
Michie, D., 'P-KP4, expert system to human being conceptual checkmate of dark ingenuity', *Computing*, 17 July 1980.
Michie, D., and Rory Johnston, *The Creative Computer*, Viking, New York, 1984.
Miller, J. G., *Living Systems*, McGraw-Hill, New York, 1978.
Mills, C. W., *The Sociological Imagination*, Oxford University Press, 1959.
Mondo, J., *Chance and Necessity*, Collins, Glasgow, 1970.
Piaget, J., *The Construction of Reality in the Child*, Basic Books, New York, 1954.
Pugh, G. E., *The Biological Origins of Human Value*, Routledge and Kegan Paul, London, 1978.
Riedl, R., *Order in Living Organisms*, Wiley, Chichester, UK, 1978.
Rose, S., *The Conscious Brain*, Weidenfeld and Nicolson, London, 1973.
Schreiber, F. R., *The Shoemaker: Anatomy of a Psychotic*, Penguin, Harmondsworth, UK, 1983.
Scriven, M., 'The mechanical concept of mind', *Mind*, No. 62, 1953, p. 246.
Simons, Geoff, *The Biology of Computer Life*, Harvester Press, Brighton, UK, 1985.
Simons, Geoff, *Is Man a Robot?*, Wiley, Chichester, 1986.
Skinner, B. F., *Beyond Freedom and Dignity*, Jonathan Cape, London, 1972.
Sloman, A., and M. Croucher, 'Why computers will have emotions', *Proceedings of the Seventh Joint Conference on Artificial Intelligence*, 1981.
Sommerhoff, G., 'The abstract characteristics of living systems', in *Systems Thinking*, ed. F. E. Emery, Vol. 1, Penguin, Harmondsworth, 1969.
Sperry, R. W., 'Consciousness, free will and personal identity', in *Brain, Behaviour and Evolution*, eds D. A. Oakley and H. C. Plotkin, Methuen, London, 1979.
Strongman, K. T., *The Psychology of Emotion*, Wiley, Chichester, UK, 1978.
Turing, A. M., 'Computing machinery and intelligence', *Mind*, Vol. 59, 1950, p. 236.
Waddington, C. H., *Tools for Thought*, Jonathan Cape, London, 1977.
Wiener, N., *Cybernetics: Control and Communication in the Animal and the Machine*, MIT Press and John Wiley, New York, 1948.
Young, J. Z., *A Model of the Brain*, Oxford University Press, 1964.
Young, J. Z., *Programs of the Brain*, Oxford University Press, 1978.

Chapter 8

Abraham, E., C. T. Seaton and S. D. Smith, 'The optical computer', *Scientific American*, February 1983, pp. 63–71.

Albinson, R., 'Biosensors – from concept to commercialisation', *Sensor Review*, January 1987, pp. 39–44.

Asimov, Isaac, and Karen A. Frenkel, *Robots: Machines in Man's Image*, Harmony Books, New York, 1985.

Biancomano, V., 'Fiber optics', *Electronic Design*, 10 July 1986, pp. 74–82.

Carlyle, R. E., 'Toward 2017', *Datamation*, 15 September 1987, pp. 142–54.

Clerman, R. J., 'Combining biology and electronics', *Data Processing*, March 1984, pp. 25–7, 30.

Davis, S. G., 'The superconductive computer in your future', *Datamation*, 15 August 1987, pp. 74–8.

Durham, T., 'Four steps to realising the sugar cube biochip', *Computing*, 25 October 1984, pp. 26–7.

Fagan, M., 'Goodbye to the silicon chip?', *Practical Computing*, August 1987, pp. 74–6.

Frude, Neil, *The Intimate Machine*, Personal Computer World/Century Publishing, London, 1983.

Gullo, K., and W. Schatz, 'The supercomputer breaks through', *Datamation*, 1 May 1988, pp. 50–63.

Hecht, J., 'Computing with light', *New Scientist*, 1 October 1987, pp. 45–8.

Main, R., 'Optical technologies – an agenda for the future', *Sensor Review*, January 1987, pp. 33–8.

Mokhoff, N., 'Parallel computer architectures of the '90s will provide solutions en masse', *Computer Design*, July 1986, pp. 64–72.

Podmore, C., and D. Faguy, 'The challenge of optical fibres', *Telecommunications Policy*, December 1986, pp. 341–51.

Reichardt, Jasia, *Robots: Fact, Fiction and Prediction*, Thames and Hudson, London, 1978.

Rifkin, G., 'On beyond silicon: a look at new semiconductor technologies', *Computerworld*, 14 April 1986, pp. 49–62.

Stableford, Brian, *Future Man: Brave New World or Genetic Nightmare*, Granada, London, 1984.

Yanchinsky, S., 'And now – the biochip', *New Scientist*, 14 January 1982, pp. 68–71.

Index

abacus 61, 63
advisers, artificial 174–177
Agriculture and Food Research
Council 145
Agrippa, Cornelius 22
Ahmes Mathematical
Papyrus 61
alchemy 21–23
algorithms 114–115, 171
Al-Jazari 47
Allen, Woody 35
Analytical Engine 65
anatomies, artificial 72, 75–91,
189–192
biological 197
see also body, warm robot
Androbot 160
animation in mythology 19, 21,
23, 29
animism 29, 30
Antikythera machine 61–62
Aphrodite, goddess 25
applications of robots *see* uses
of robots
Apprentice robot 129, 140
Aquinas, Thomas 51, 52
Archimedes 44
Aristotle 55
arms, artificial 76, 83–85
Arnold of Villanova 22
Ars Magna 62
artificial intelligence
(AI) 60–61, 97, 98–99,
105–106, 115, 118,
122–124, 138–139, 145,
147, 156, 168–177, 193,
204–206, 207, 208
Asimov, Isaac 33–34, 36–37,
71, 169, 181–182
Ask, artificial man 18
Assyro-Babylonian
mythology 18
Athene, goddess 26
attitudes, robot impact
on 194–196
Australia 148
Australian mythology 19, 20
automata, early 41–68
Automatic Sequence
Controlled Calculator
(ASCC) 66
automotos 41
autonomy 41–42, 94, 96–97,
98, 106
see also free will
Aviram, Arieh 207

Babbage, Charles 64–66
Babbage, Henry 66
Bacon, Roger 51, 52
Bandai Company 59
barmen, robot 128
Basile, Giambattista 25
Bertran, Charles 57
Bhagavad Gita 17
biochips 206–207, 208
Black Hactin, god 17
body, warm robot 167–168, 189
*Book of Knowledge of
Ingenious Mechanical
Devices* 47
brain, mammalian 98–99, 199
robot 98–99
see also artificial intelligence
Brihatkathasaritsagara 39
Bristol University 145
British Leyland 139
British Oxygen Company 139
British Robot Association
(BRA) 71, 129
Brunel University 113
Bruno, Giordano 62
Buddhism 30, 50
Bujanska, Maria 183
Bullmann, Hans 52
butcher, robot 129, 145
Butler, Samuel 41

Cabbala 22, 23
cabbalism 22–23
calculators, mechanical 61–66
California University 150, 151
Capek, Karel 33
Carnegie Mellon
University 13, 123, 152,
157, 161, 190
Carron Engineering 143
Celsus 45
central processing unit
(CPU) 118, 126
Charles Stark Draper
Laboratory 108
Chernobyl 157
China 45, 61, 48–49, 105
Chinese mythology 18
Chomsky, Noam 200
Cincinnati Milicron 129, 130,
140
Citroën 140–141
cleaner, robot 13
clepsydras 46
Clifford, Paul 190
clocks 51–52, 54, 62–63
connectionist circuits *see*
neural computing
CONSIGHT system 117
control systems 72, 95–104,
132, 143
Coppelia theme 34, 177
Cornell University 13
Cranfield Institute 79
creation myths 16–20
creativity 201–202, 203
Cruchet, Marie 57
'crucial convergence' 167–168,
169, 176, 208
Ctesibus 44
cybernetics 170–171, 196,
198–199, 201, 202, 204
Cyprus 25

Daedala 45
Daedalus 18, 45
Dartmouth College 169
Dawkins, Richard 195–196
De Generatione Rerum 22
Democritus 194
Denning Mobile Robotics 155
Department of Trade and
Industry, UK 148
Depero, Fortunato 36
Der Sandmann 32–33, 177
Descartes, René 51
DeVilbiss Trallfa
robot 142–143
Devol, George 70
Dickens, Charles 41
Difference Engine 65
dolls, 'living' 56–60, 178
domestic robots 160–162, 191
Droz, Pierre and Henri-
Louis 53

ears, artificial 122–124
East India Company 49
Edison, Thomas 57, 66–67,
178–179
Egypt, ancient 46
Egyptian mythology 17
Eleazar of Worms 23
electronics 66–67
see also neural computing;
semiconductors
Elijah of Chelm 23
Embla, artificial woman 18
end effectors 83
see also hands, artificial
Engelberger, Joseph 29, 70–71,
152, 161
entertainment, robots
in 163–164
Epaulard robot vehicle 158
Epic of Gesar of Ling 21
Epic of Gilgamesh 17
Euclid 44
Euphonia talking machine 54
experts, artificial 172–177
expert systems 172–177, 207
eyes, artificial 75–76, 100–101,
113–122, 138, 141

Faber, Joseph 54
factory, unmanned 129, 132, 144
feet, artificial 162
Fiat 138, 140, 142
films, robot 34–36, 185–187
fingers, artificial 76, 78–79, 80–83, 109, 179
fire-fighting robot 125
Fluid Technology Laboratory 81
food processing robot 147
Ford 140
France 109, 140–141, 150
Frankenstein 27, 28, 35
Frankenstein, Victor 15
free will 31, 200–201, 202–203
fruit-harvesting robot 147
future trends 204–208
fuzzy logic 206

gaits 89
Galatea 25
Galileo 28–29, 105
game-playing robot 163
General Electric 86, 138
General Motors 71–72, 129, 135
generations, robot 73
Germany 139–140
gods, as first robot engineers 15–20
golem 21, 23–24, 28, 34, 35
Greek mythology 18, 26–28, 39
Greeks, ancient 42–43, 46, 61, 62, 194
grippers *see* hands, artificial
Gulf War (1991) 153, 176
Gulliver's Travels 62

Hall Automation 139
hands, artificial 74–84
hand, Utah/MIT 82
Hasbro-Bradley Company 59
Hather, goddess 17
health care, robots in 148–152
hearing, artificial *see* ears, artificial
Helen O'Loy 33, 181
Hephaestus, god 19, 26, 28, 35, 39
Hercules, god 27
Hermes, god 26–27
Hero 44
Herodotus 85
Heseltine, Michael 195
heuristics 170, 171
Hoenir, god 18
Hoffmann, E. T. A. 33
Hofstadter, Douglas 168
hopping robots ('hoppers') 87–88, 90
Hosokawa, Yorinao 48
Hughes Aircraft 117–118
Hull University 76–77
hydraulics 92
hydraulos 92
Hypatia 28

Ibn Al-Razzaz Al-Razzaz 47
Icarus 45
Ihnatowicz, Edward 42
Iliad 71
Imperial College 80, 145, 150, 151
India 46, 49–50, 85, 105, 194
Indiana University 168
Indian (North American) mythology 17

industrial robots 29–30, 72, 84, 129–144, 164–165
integrated circuits 171
  see also semiconductors
International Harvester 134
Islamic culture 46
Italy 140

'Jacks' (clocks) 51–52
Jacobson, Jewna 64
Jacquard, Joseph 64
Japan 29–30, 31, 37–38, 47–48, 50–51, 58–60, 70, 71, 77, 79, 88, 117, 132, 137, 147, 149, 150, 157, 160, 163–164, 189, 190, 198, 207
Japanese Industrial Robot Association (JIRA) 71
'Jaquemarts' (clocks) 51–52
Jewish mythology 21–22, 23–24, 34
Judaeo-Christian tradition 17, 28–29, 30–31
  see also Jewish mythology

Kamata, Satoshi 196
Kanazawa Industrial University 128
karakuri (automata) 48, 50, 56
karakuri masters 50–51, 53
Karakurizui 48
Keele University 13
Kepler, Johannes 63
Khnum, god 17
khwai shuh, skill of 21
Kitahara, Teruhisa 59
knowledge management 174, 192
KUKA robots 139, 141–142

Lang, Fritz 35, 185
languages, robot 102
Larrson, Ove 76
'laws of thought' 61
legs, artificial 85–91
Leibniz, G. W. 62, 64
Lem, Stanislaw 183
Lengyel, Jed 13
Leonardo da Vinci 52
Lester del Rey 33, 181
L'Eve Nouvelle (L'Eve Future) 32, 178
librarian, robot 128
Locoman robot 84
Lodur, god 18
logic, binary 67
  see also fuzzy logic
London University 169
Louisiana State University 147
Lovelace, Ada 65
lovers, robot 187–192
Low, Rabbi Judah 23, 30–31
Lucian 194
Lucretius 194
Lull, Ramon 62

MacLeod, Sheila 183
Maezel, Leonard 56
Magnus, Albertus 51, 52
Magnus, Simon 22
Mahabharata 17, 39
Mami, goddess 18
Maori mythology 17
Marduk, god 18
Martin, James 205–206
Massachusetts Institute of Technology (MIT) 115
Matsubara, Sueo 31
McCarthy, John 169, 170

McCulloch, Warren 171
McDonnell Douglas plant 132
McEnroe, John 195
Mecanotron Corporation 111
medical aspects 188–189, 193
  see also prostheses; health care, robots in
Melanesian mythology 17
'Meldog' guide dog 149
Méliès, Georges 34–35
memory, artificial 100
mental disease, computer 208
Metal Castings Company 97
Metropolis 35, 185
Microbot 159
microcomputers 97–98, 124
microprocessors 97, 103
Midgard (or Mana-heim) 18
milkmaid, robot 146–147
Minsky, Marvin 169, 171
mobility, robot 85–91
Monod, Jacques 198
Mori, Masahiro 31
Morland, Samuel 64
motive forces *see* power units
motors 88, 91–92, 154
Mount Caucasus 27
Mount Olympus 26
muscles, artificial 76–77, 179
music-playing robot 163–164
mythology 15–40

Nammu, goddess 16
Nanyang Technological Institute 123
Napier, John 63
National Toy Company 57
neural computing 171–172, 204
Newcastle University 112
Newell, Allen 169
New York University 82
Nissan 133
Norse mythology 18, 85
Nottingham University 129
nuclear plant, robots in 157–158

O'Brien, Edna 182
occultism *see* cabbalism
Oceania mythology 18–20
Odex robot 93
Odin, god 18
Olympic Ode 45
Omnibot robots 57
Omnigripper 80
Ono, Benkichi 51
optical circuits 77, 207

Pandora, artificial woman 18, 26–27
Paolozzi, Eduardo 36
Papert, Seymour 159, 171
Paracelsus 22, 23, 28
parallel circuits 205–206, 208
  see also neural computing
Paré, Ambroise 78, 86
Pascal, Blaise 63–64
Pascaline calculator 63–64
patent, early robot 70
Pausanias 44
Pentamerone 25
perceptrons 171
Petty, Charles 154
Peugeot 140, 141
Philon 44
'pick and place' robots 69–70, 71
Pindar 45
Pisa University 77
Pitts, Walter 171

planetary robot 13
Plato 44
pneumatics 92, 95–96
power units 72–73, 79, 91–92
Prab robots 134, 143
Pragma robot 137, 138
prisons, robots in *see* sentry, robot
programmability 71, 72
programming, human 199–201
robot 101–103, 124
programs, vision 115–116, 170
Prometheus, god 18, 24, 26–28
Propoetides, the 25
prostheses 78, 85–86, 88–89, 148, 150–151, 188–189, 193
Prowler robot 153–154
PUMA robot 129, 135, 148, 150, 161
Pygmalion 24, 25–26, 34

Quasar Industries 160

receptionist, robot 129
Reichert, Mark 13
religion 28–32, 34
see also Judaeo-Christian tradition
remote centre compliance (RCC) 108
Rhino Robots 159
Rhode Island University 117, 121
Ricci, Matthew 48
Rig-Veda 85
Robart project 156
Robodoc 151
Robogate system 140
'robot', word origin 33
Robotarm 84
robot ballet 36
robot defined 71–72, 74
Robotic Industries Association (RIA) 71
Rolinx 143
Romans, ancient 43, 46, 61
ROSA robot 93
Rossum's Universal Robots (RUR) 33
Russell, Bertrand 29

Saab-Scania Company 131–139
Sakamoto, Masaki 37
Sanskrit mythology 17–18
SCARA design 137
Schickard, William 63
Science and Engineering Research Council 145
Seidon, Philip 207
semiconductors 67, 97, 98, 100, 106, 117–118, 206, 207
Senkereb Tablet 61
senses, artificial 100–101, 104–127, 145, 149, 155, 190, 206
sensors *see* senses, artificial
sentry, robot 125, 128, 155–156
SERIE robot 138
set-up, physical 95
sex aids 188, 193
sheep-shearer, robot 129, 146
Shelley, Mary 15, 27, 208
Shintoism 29–30
Sigma robot 137
silicon *see* semiconductors
Simon, Herbert 169
skin, artificial 108–109, 111, 179
Skywasher robot 162
Sloman, Aaron 169
SMART robot 138
smell, artificial 124, 125
speech, artificial 54, 56, 57
speech recognition 122, 123
Spider Woman 17
spine, artificial 76
Stanford Research Institute 136
Steiner, Jules Nicholas 57
Sumatran mythology 19
Sumerian mythology 16–17
superconductivity 206
surgery, robot 149–152
surrogate people 13–14, 25, 166–193, 208
see also uses of robots
Sussex University 169
Swahili mythology 17
Sweden 70, 81, 131, 139, 148
Switzerland 43

tactile sensors *see* touch, artificial
Takeda, Omi 48
Talmud 17, 21, 85
Talus, artificial giant 18, 35, 39
Tanaka, Hisashige 50
Tanaka Seizojo Company 50–51
taste, artificial 125
teaching robots 101–102
Ten Principles of Robot Law 37–38
Tezuka, Osamu 37
The Cyberiad 183
Theroude, Alexandre Nicholas 56
The Senster, cybernetic sculpture 42
The Uncommercial Traveller 41
thought, artificial *see* artificial intelligence (AI)
Three Laws of Robotics (Asimov) 33, 36–37
Three Mile Island 157
Thring, M. W. 74, 93, 160, 169
Tilottama (artificial woman) 18
Tin Toy Museum, Yokohama 58–59
Tippoo's Tiger 49
TI Tubes Company 131, 133
Tomy Company 60
Topo robot 160
Topping, Mike 13
Torres y Quevado, Leonardo 55
torso, artificial 75
Toshiba 51, 133
touch, artificial 107–113
Toyota 132
training, robots in 159–160
transistors 67, 206, 207
transphasor 207
Treatise on Pneumatics 44
tree-spraying robot 147
Turing, Alan 170
Turing Institute 121
Turtle robot 159
types, robot 73–75

undersea robots 158
Unger, Felix 189
Unimate robots 129, 131, 132–133, 139, 143–144
Unimation 70, 71, 129, 139
United Kingdom 134, 142, 143, 145, 169, 188–189
United States 59, 69, 70, 88, 93, 117, 125, 129, 131, 132, 133, 151, 153–154, 155, 160, 161, 168, 207
uses of robots 86–87, 128–165
barmen 128
butchers 129, 145
domestic 160–162, 191
entertainment 163–164
fire-fighting 165
food-processing 147
fruit-harvesting 147
game-playing 163
health care 148–152
industry 29–30, 72, 84, 129–144, 164–165
librarians 128
milkmaids 146–147
music-playing 163–164
nuclear plant 157–158
outer space 158–159
receptionists 129
sentries 125,128, 155–156
sheep-shearers 129
surgery 149–152
training 159
tree sprayers 147
undersea 158
waiters 147
war makers 153–155
window cleaners 162
see also surrogate people
USSR 86

Vale, Mark 69, 70
valves, thermionic glass 66
Vaucanson, Jacques de 52, 53
Ve, god 18
Ventaron, Jean Mathieu de 49
Vili, god 18
Villiers de l'Isle Adam 32, 178
vision, artificial *see* eyes, artificial
Visvakarman, god 17, 18
voice recognition 122–124
Volvo 139, 140
von Kempelen, Wolfgang 53, 54
von Knauss, Friedrich 52
von Neumann, John 170

Wabot robots 88, 163–164
waiters, robot 147
war, robots in 153–155
Waseda University 79, 88, 163–164
Watson, Thomas 70
Weizenbaum, Joseph 171
Westinghouse Corporation 136
Wiener, Norbert 30–31, 55, 169–170, 196, 198, 204
Wilde, Oscar 26
window-cleaning robot 164
Wisard system 113
Wisconsin University 86, 88
World Health Organization (WHO) 149

Xavier, Francis 48

Yamanashi University 137
Yokoyama, Mitsuteru 38
Yokoyama, Ryuichi 37
Yugoslavia 109
Yu-ti, god 18

Zeus, god 26, 27